Science
Library

Volume

22

Milton H. Saier, Jr.
Charles D. Stiles

Molecular Dynamics in Biological Membranes

Springer-Verlag
New York
Heidelberg
Berlin
1975

Milton H. Saier, Jr.
Department of Biology
John Muir College
University of California at San Diego
La Jolla, California 92037

Charles D. Stiles
Department of Biology
John Muir College
University of California at San Diego
La Jolla, California 92037

Library of Congress Cataloging in Publication Data

Saier, Milton H. 1941-
    Molecular dynamics in biological membranes.

    (Heidelberg science library; v. 22)
    Includes index.
    1. Membranes (Biology) 2. Molecular biology.
I. Stiles, Charles D., joint author.   II. Title.
III. Series. [DNLM: 1. Molecular biology. 2. Cell
membrane—Physiology. 3. Perception. 4. Metabolism.
5. Biological transport. QH601 S132m 1976]
QH601.S24   574.8'75   75-12923

ISBN 0-387-90142-6   Springer-Verlag   New York

ISBN 3-540-90142-6   Springer-Verlag   Berlin Heidelberg

to Charles Frisbie
    Margaret Rowell
    Clinton Ballou
    Saul Roseman

Acknowledgments

We would like to thank our colleagues and friends who contributed to the formulation of this volume:

Clint Ballou
Mark Bashor
Bert Ely
George Fortes
Clem Furlong
Helen Hansma
John Judice
Howard Kutchai
Jack Kyte
Lola Reid

Jeanne Saier
Lucelia Saier
Birgit Satir
Mel Simon
Ruth Ann Stiles
Jon Singer
Nick Spitzer
Walther Stoeckenius
Juan Yguerabide

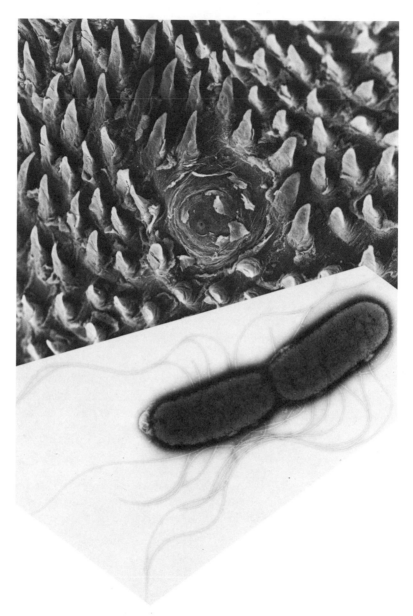

TOP: Scanning electron micrograph of a taste bud surrounded by filiform papillae on the tongue of a mouse.

Courtesy of Dr. Jean-Paul Revel, Department of Biology, California Institute of Technology.

BOTTOM: Electron micrograph of a motile *Escherichia coli* cell undergoing division.

Courtesy of Dr. Melvin Simon, Department of Biology, The University of California at San Diego.

Preface

There is something fascinating about science. One gets such wholesale returns of conjecture out of such a trifling investment of fact.

Mark Twain

The recent explosion in our knowledge of basis physiologic processes, molecular biology, and genetic regulatory mechanisms has resulted, in large measure, from a single conceptual advance: the realization that, at the molecular level, evolutionarily divergent organisms are more similar than different. Thus, in *Escherichia coli* and *Homo sapiens*, the enzymatic pathways for the utilization of galactose and glucose are the same, although more than a single sequence of enzymatic reactions can lead to the utilization of either sugar. Also, extensive studies have revealed the essential universality of the genetic code, the mechanism of deoxyribonucleic acid (DNA) replication, and the processes by which genetic information is transcribed to ribonucleic acid (RNA) and RNA is translated into protein. A detailed comparative examination of any one area of biologic interest, of course, reveals differences among phylogenetically distinct organisms. In prokaryotic organisms protein synthesis is initiated with *N*-formyl methionyl-transfer RNA (tRNA), whereas methionyl tRNA serves this function in the cytoplasm of the eukaryote. Mechanistic differences may have evolved to accommodate the differing degrees of complexity of cellular construction or to coordinate functions of differentiated cells in a multicellular organism. Yet, we must realize that the basic, life-endowing molecular processes had to exist prior to extensive evolutionary divergence—before the appearance of two distinct cell types.

Consequently, we should expect that these processes are governed by the same principles and that even the molecular details will frequently have been conserved throughout evolutionary history.

The applicability of this unifying maxim to membrane biology is not yet as clearly recognized as it is in more extensively understood areas of biology. Thus, although it is generally accepted that membrane structure and biogenic mechanisms are likely to be universal (Chapters 2 and 3), fewer biologists would acknowledge the same for subjects such as exo- and endocytosis (Chapter 4), transmembrane solute transport (Chapter 5), chemical and energy reception (Chapters 6 and 7), and metabolic regulation (Chapter 8). Nor do most biologists believe that the molecular principles governing cellular recognition and social behavior (Chapter 9) will be found to be generally applicable across phylogenetic lines or that bacterial physiologic studies will provide a valid guide to human pathology (Chapter 10). In many cases, these doubts cannot be easily dispelled: too few facts have yet accumulated to allow generalization. But, in those instances in which sufficient knowledge about a membrane-associated process is available, the basic evolutionary conservatism noted above appears to be substantiated.

The present monograph is concerned primarily with the application of this fundamental precept to specific areas of membrane biology. In attempting to illustrate these principles, we will wander to the edges of (and beyond) the frontiers of our scientific knowledge. We will examine biologic systems from a phenomenologic standpoint, and, when possible, scrutinize these systems at the molecular level. In a few instances we will be preoccupied with biologic aesthetics, noting that the concept of a biologic process may appeal sensually to an individual at several different levels. Although the pages that follow are meant to familiarize the reader with a few of the rapidly advancing areas of membrane biology, extensive reference to experimental detail has been intentionally omitted in order to maximize conceptual recognition of the underlying principles governing modern membrane research. Only when a knowledge of experimental protocol is essential to an understanding of the process under discussion, will this information be presented. Selected references at the end of each chapter are provided to allow the reader to pursue a subject in greater depth.

Contents

10

Role of the Plasma Membrane in Growth Regulation and Neoplasia 114

1 Introduction: Cell Structure and Function

The harmonious co-operation of all beings arose, not from the orders of a superior authority external to themselves, but from the fact that they were all parts in a hierarchy of wholes forming a cosmic pattern, and what they obeyed were the internal dictates of their own natures.

Chung Tzu

The fundamental unit of structure in biology is the living cell. Regardless of whether an organism is single celled or multicellular, it must carry out the same basic functions in order to survive, both as an individual and as a species. These functions include acquisition of nutrients and energy sources, disposal of unusable and toxic materials, reproduction, locomotion, and interaction with components in the environment. All of these functions require coordination both for short-range activities, such as sensation, and for long-range activities, such as growth. A comparison of organisms across phylogenetic lines reveals similarities, which have led to a formulation of the *unity principle* as applied to biology:

There is an underlying simplicity in the structural and functional organization of a cell from which we can infer that all living organisms have a common ancestor. As a consequence of this common ancestry, an essential process which confers upon an organism one of its life-endowing attributes will be similar in principle, throughout the living world. Differences must have arisen as a result of evolutionary divergence.

All cells are characterized by a plasma membrane, which encapsulates the cytoplasm and creates internal compartments in which essential functions are carried out. Two basic types of cells have been characterized: prokaryotes and eukaryotes. Prokaryotic cells are characteristically small and possess minimal intracellular structure. In many such cells, distinct membrane-bounded organelles are absent. By contrast, the eukaryotic cell is much larger and contains numerous intracellular organelles of widely differing structure and function (Figure 1.1), each of which is specialized for one function: digestion, respiration, biosynthesis, or secretion (Chapter 4). Since both bacterial and eukaryotic cells must carry out the same essential processes to live and reproduce, it is clear that these differences in cell structure do not indicate divergent modes of life but merely the presence or absence of compartments specifically designed to fulfill essential functions. In the eukaryotic cell, each process is performed in a spatially isolated domain, whereas these processes operate largely within a single compartment in the prokaryotic cell. Thus, whereas the prokaryotic cell may at first glance appear less complicated than the cells of higher organisms, the functional dissection of the latter may prove simpler.

To exemplify the unity principle, one can compare the process of energy generation in eukaryotic and prokaryotic cells. Among eukaryotic organelles is the mitochondrion, the organelle of respiration (Figure 1.1). Mitochondria possess outer and inner membranes of entirely different composition. By virtue of this structural feature, the mitochondrion itself can be thought of as a multi-compartmental structure: its four domains include the inner and outer membranes, the matrix between these two membranes, and the inner matrix. Different enzymes are associated with each domain. For example, only in the outer membrane does one find the enzyme, monoamine oxidase, whereas the protein components of the electron transport chain and the oxidative phosphorylation system are localized in the inner membrane. One finds adenylate kinase in the outer matrix, but the enzymes that oxidize fatty acids, amino acids, and di- and tricarboxylic acids appear exclusively in the inner matrix. Interestingly, the inner matrix also contains deoxyribonucleic acid (DNA) and all of the machinery necessary to transcribe and translate the nucleotide sequences of this nucleic acid into protein. This observation has led to the suggestion that mitochondria may be capable of self-replication, to a degree, and that these eukaryotic organelles may have evolved from a prokaryotic cell. In fact, the structural and functional features of the mitochondrion are remarkably similar to those of Gram-negative bacterial cells. In both, the outer membranes are permeable to protons and other small ions, whereas the inner membranes are

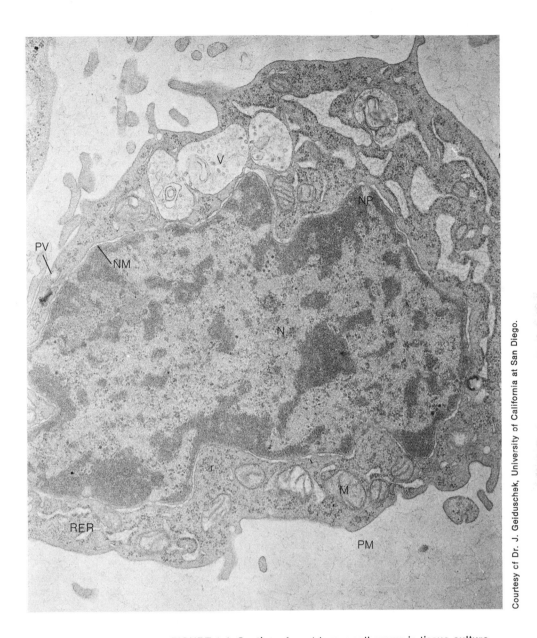

FIGURE 1.1. Section of a rat tumor cell grown in tissue culture. The electron micrograph reveals an extraordinarily large nucleus (N) surrounded by a double nuclear membrane (NM), in which numerous nuclear pores (NP) can be observed. In the cytoplasm are found mitochondria (M) with cristae and outer membranes, numerous vacuoles (V), ribosomes (r), and rough endoplasmic reticular membranes (RER). The unit plasma membrane (PM), which surrounds the cell may invaginate with the formation of a phagocytic or pinocytic vesicle (PV). Magnification ×7300.

not. In both, the inner membranes possess proteins that facilitate the transport of small molecules across the hydrophobic barrier. In both, the inner membranes house the enzymatic machinery involved in the interconversion of different forms of energy (Chapter 5).

When a mitochondrial or bacterial suspension is provided with a supply of an oxidizable substrate, such as succinate in the presence of air, three observations can be made: first, gradients of ions across the inner membrane are generated, and these give rise to a transmembrane electric potential; second, ATP is synthesized; third, the inner membrane undergoes changes in conformation. Thus, energy can presumably be stored in three different forms: in the electroosmotic gradients of ions, in the chemical pyrophosphate bond of ATP, and in the conformational states of inner membrane proteins. As will be discussed in Chapter 5, these different energy forms may be utilized to drive the active accumulation of nutrients and other beneficial molecules into the cytoplasm. The extrusion of toxic materials from the cell is similarly accomplished by energy-dependent transport systems which function with outwardly directed polarity.

Considerable evidence suggests that some eukaryotic organelles have evolved by degeneration of an endosymbiotic prokaryote within a pre-eukaryotic cell. Thus, mitochondria may have evolved from aerobic eubacteria, chloroplasts from blue-green bacteria, and, possibly eukaryotic flagella and cilia from bacterial ancestors resembling spirochetes. These possibilities are strengthened by the observation that present-day bacteria are known to propagate within the cytoplasm of eukaryotic organisms in a relationship that is mutually beneficial to the two organisms. In view of the probable evolutionary relationships of bacteria to eukaryotic organelles, the study of bacterial physiology can be expected to shed light on corresponding processes in eukaryotic systems. Similarly, studies with eukaryotic organelles should enhance our understanding of prokaryotic physiology.

Surface components of cells of the two major subdivisions of life show characteristic properties. These are compared schematically in Figures 1.2 and 1.3, to illustrate the essential features of bacterial cell envelopes and those of eukaryotes, respectively. Each of the envelopes depicted is characterized by a plasma membrane and outer layers, which may include a rigid cell wall. The macromolecules that comprise the envelope provide the cell with its identity, its fingerprint. For example, the glycoproteins and glycolipids on the surfaces of animal cells include the blood group substances, the histocompatibility and transplanta-

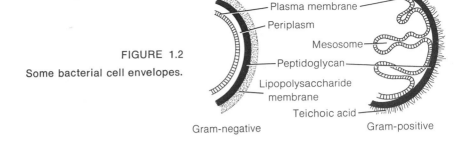

FIGURE 1.2

Some bacterial cell envelopes.

surface macromolecules allow for cellular recognition, and they provide the structural basis for tissue formation in multicellular organisms.

Bacteria similarly possess integral outer membrane constituents, which confer upon a species its identity. These molecules include the cell surface lipopolysaccharides (endotoxin), envelope proteins, and carbohydrate-containing moieties of the rigid cell wall in the Gram-negative bacterium. The Gram-positive bacterium possesses only one membrane, the cytoplasmic membrane, and a rigid cell wall. Although the outer lipopolysaccharide membrane is lacking, carbohydrate-rich molecules, the teichoic acids, are covalently linked to the rigid peptidoglycan layer, and these species-specific molecules are partially responsible for the antigenic properties of the cell.

A cell that can be recognized but that cannot recognize

FIGURE 1.3. Some eukaryotic cell envelopes.

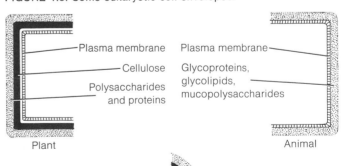

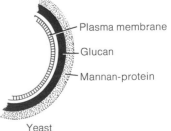

its neighbors or constituents in its environment would clearly be a social failure, and it would not be likely to have evolved to the present millennium. Thus, one finds on the surfaces of animal cells a variety of receptors: chemoreceptors, which inform the cell of useful and toxic chemicals in the environment (Chapter 6); energy receptors, which relate the presence of light, heat, electricity, pressure, and gravity (Chapter 7); macromolecular receptors, which recognize hormones and certain drugs (Chapter 8); and, finally, surface components, which function to recognize and bind viruses, bacteria, and other animal cells (Chapters 9 and 10).

Similar receptors are found on the surfaces of bacterial cells. These simplest of living organisms swim toward sweet and spicy compounds while avoiding bitter or sour compounds (Chapter 6). They seek light and warmth and respond to electricity, pressure, and gravity (Chapter 7). Bacterial surface receptors regulate cellular metabolic processes as do hormone receptors in animal cells (Chapter 8). Moreover, microorganisms have evolved surface macromolecules that function as receptors for viruses, dissimilar bacterial species, and bacteria of the same species (Chapters 9 and 10). The latter observation implies the presence of specialized adhesive structures analogous to those that must exist on the surfaces of animal cells.

Like most of us, cell envelopes possess two sides: Those macromolecules that confer upon a cell its sense of identity and allow communication with the environment are generally localized on its external side. By contrast, membrane proteins involved in intracellular metabolism are found on or within the inner surface of the membrane. Included among such enzymes are adenylate cyclase, which synthesizes a cytoplasmic messenger of hormone action, cyclic AMP (Chapter 8), and membrane-bound ATPases, which appear to function in the transmembrane transport of small ions (Chapter 5). These proteins are found in both prokaryotic and eukaryotic plasma membranes, where they sense external stimuli (through receptor proteins) and regulate the intracellular environment, thereby influencing intracellular metabolism. Current evidence favors the notion that the cellular transmission mechanisms that link stimulus to response in simple bacterial and eukaryotic cells are similar to those in animal cells, which are highly differentiated for electric impulse propagation or endocrine control (Chapters 7 and 8). Thus, a useful model system for investigating some aspects of neurophysiologic behavior or endocrine regulation may be *Escherichia coli*. These observations suggest that mechanisms responsible for complex membrane phenomena will be found to be universal throughout the living kingdom; principles established with one biologic system will apply to others. In conclusion, the unity principle is in; the narrow view is out!

Selected References

Loewy, A. G. and P. Siekevitz. *Cell Structure and Function.* 2nd Ed., Holt, Rinehart and Winston, Inc., New York, 1969.

Margulis, L. *Origin of Eukaryotic Cells. Evidence and Research Implications for a Theory of the Origin and Evolution of Microbial, Plant and Animal Cells on the Precambrian Earth.* Yale University Press, New Haven, Conn., 1970.

Novikoff, A. B. and E. Holtzman. *Cells and Organelles.* Holt, Rinehart and Winston, Inc., New York, 1970.

Paecht-Horowitz, M. The origin of life. *Angew. Chem.* (Inter. Ed.), *12*:349 (1973).

Society for General Microbiology, *Evolution in the Microbial World, 24th Symposium of the Society for General Microbiology.* Cambridge University Press, London, 1974.

Stanier, R. Y., M. Deudoroff, and E. A. Adelberg (eds.). *The Microbial World.* 3rd Ed., Prentice-Hall, Inc., Englewood Cliffs, N.J., 1970.

Starr, M. P. "Bacterial diversity: The natural history of selected morphologically unusual bacteria," in *Annual Review of Microbiology, Vol. 19* (C. E. Clifton, S. Raffel and M. P. Starr, eds.). Annual Reviews, Inc., Palo Alto, Ca. 1965, p. 407.

Trinkaus, J. P. *Cells Into Organs.* Prentice-Hall, Inc., Englewood Cliffs, N.J., 1969.

Wittaker, R. H. New Concept of kingdoms of organisms. *Science, 163*:150 (1969).

2 Constituents of Biological Membranes

There is no love in the little houses.
Love is outside, in that storm,
Playing with mountains
With long fingers of light.
I might show you love
But you must follow me
Out of the little houses.

Charles Frisbie

An understanding of any one process that occurs within the matrix of a biological membrane requires some knowledge of the constituent elements that comprise this structure. Moreover, information regarding the nature of permissible interactions that occur between membrane constituents may lead to insight into the functioning of membrane-associated multicomponent complexes. For these reasons, we initiate our discussion of biological membranes with a consideration of the nature of the structural and catalytic elements that comprise these macromolecular complexes. Just as the functioning of a cell is dependent on its structure, structure must have evolved to accommodate all essential functions.

Membrane Isolation The difficulty with which a pure membrane preparation can be isolated depends on the complexity of the cell from which it is derived. From the standpoint of the membrane biologist, the mammalian red blood cell is among the simplest of living cells, being, essentially, a sack of soluble hemoglobin. If the cell is subjected to hypoosmotic conditions in the absence of stabilizing divalent cations, water will rush into the cell, and the increased intracellular volume will cause the membrane to rupture. This procedure is referred to as *hypotonic* or *osmotic shock*. During osmotic shock treatment, soluble proteins and a fraction of the membrane-associated protein escape from the cell interior. Subsequently, the membrane reseals when brought back to isotonic conditions with the formation of a membrane "ghost" devoid of soluble constituents. Electron microscopic examination of ghost preparations has revealed that the size of these vesicles is comparable to that of the cells from which they are derived, suggesting that the procedure does not cause appreciable membrane fragmentation. Moreover, the sacks are "right side out." The asymmetry of the plasma membrane is retained in the ghosts so that essentially all of the carbohydrate moieties of the membrane glycoproteins and glycolipids are localized on the external surface.

A similar procedure has been applied to bacteria. But since the rigid cell wall prevents osmotic swelling, this layer must first be disrupted by exposure of the organism to an enzyme that hydrolytically cleaves the covalent bonds of the wall polymers. The enzyme most frequently applied to bacteria for this purpose is egg white lysozyme. The *spheroplasts*, which result after treatment of bacterial cells with this enzyme, are spherical membrane sacks, which, unlike the bacteria from which they were derived, are sensitive to the osmotic shock procedure described above (Figure 2.1).

Electron microscopic examination of membrane preparations derived from osmotically shocked, Gram-negative bacterial spheroplasts has revealed closed, double-membrane vesicles in which the outer, but not the inner, membrane is

FIGURE 2.1. Procedures for fractionating the Gram-negative bacterial cell.

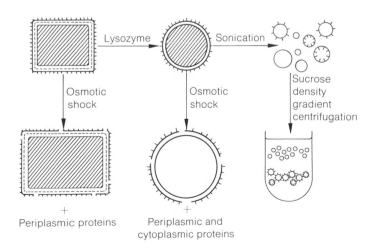

frequently fragmented (Figure 2.1). The vesicles are of a size comparable to that of the intact cell, and the orientation of the membrane constituents is generally the same as in the intact bacterium. In these preparations, only the inner cytoplasmic membrane provides an osmotic barrier. The outer lipopolysaccharide membrane is permeable to ions and small solute molecules.

Both bacterial and animal membrane vesicles provide simple systems for the characterization of a variety of membrane phenomena, such as solute transport, electron flow, and energy interconversion. Biosynthetic reactions and receptor-dependent processes have also been studied employing these systems. Since several of these processes occur in the virtual absence of soluble cell constituents, the necessary enzymatic machinery must be entirely associated with the membrane fraction.

As noted in Chapter 1, the bacterial cell, like the mitochondrion, is compartmentalized. This fact was first revealed by electron microscopic and biochemical studies, which allowed identification of the constituents found in each of the bacterial compartments (Figure 2.1). Osmotic shock of an intact Gram-negative bacterium specifically releases the proteins localized within the *periplasmic space* (between the inner and outer membranes) because the outer membrane is external to the sieve-like peptidoglycan cell wall and, therefore, will swell under hypotonic conditions. Rupture of the inner membrane following lysozyme treatment results in release of both periplasmic and cytoplasmic soluble proteins. Thus, soluble constituents of the bacterial cell can be localized by the criterion of release by osmotic shock.

Since the inner and outer membranes were found to differ with respect to composition and density, they could be separated by a procedure that allowed fragmentation of these structures. Treatment of bacterial spheroplasts with high frequency sound (*sonication*) accomplished this goal, and, since the membrane fragments resealed in a random fashion, small vesicles that were either right side out or wrong side out resulted. Regardless of orientation, the vesicles derived from the two membranes could be separated by high speed centrifugation in a continuous gradient of aqueous sucrose, in which the sucrose concentration, and therefore the density of the solution, was highest at the bottom of the centrifuge tube. In such a *sucrose density gradient*, the membranes reached equilibrium at positions in the gradient that corresponded to their respective densities (Figure 2.1).

Specific bacterial membrane proteins are generally found to be located exclusively in either the inner or outer membrane, although a few are present in both structures. Moreover, a soluble protein is usually localized in the cytoplasm, the periplasm, or the extracellular fluid (Table 2.1). The fact that individual proteins are compartmentalized suggests that complex secretory mechanisms are operative in bacteria: a soluble protein may pass through none, one, or two membranes before reaching its site of action. A lipophilic protein may be inserted into either the inner or outer membrane and be exposed at either of the two surfaces of one of these bilayers. The mechanisms that ensure proper subcellular localization of proteins are still unknown.

Eukaryotic cells exhibit a degree of structural complexity that far exceeds that of the bacterium (see Chapter 1). Plasma and cytoplasmic membrane structures must be distinguished and separated before biochemical analyses of animal cell membranes are possible. Separation of membrane fractions initially requires the disruption of the plasma membrane under conditions that leave subcellular organelles intact. A common procedure involves mild homogenization in a slightly hypotonic solution. In order to minimize membrane aggregation and disruption, the osmolarity is controlled with sucrose, and the solution is buffered near neutrality.

A variety of techniques has been developed for the fractionation of subcellular organelles from disrupted animal cells. The most frequently used procedures are differential centrifugation, which sediments organelles on the basis of size, and equilibrium density gradient centrifugation, which utilizes the different densities of the intact organelles. Figure 2.2 illustrates the technique of differential sedimentation. Nuclei, being the largest of the eukaryotic organelles, sediment to the bottom of the centrifuge tube at the low-

Table 2.1 Subcellular Location of Proteins in Gram-Negative Bacteria

Extracellular	Inner Cytoplasmic Membrane
Proteases	Electron transfer chain
Lipases	Proton-translocating ATPase
Carbohydrases	Some transport proteins
Nucleases	Lipid and cell envelope biosynthetic enzymes
Outer Lipopolysaccharide Membrane	**Cytoplasm**
Receptor proteins for viruses and bactericidal agents	Enzymes that catalyze the synthesis and degradation of soluble substrates
Phospholipase A and lysophospholipase	DNA
	RNA
Periplasm	Ribosomes
Solute-binding proteins involved in transmembrane transport and chemotaxis	Enzymes involved in DNA replication, transcription, translation, etc.
Phosphatases and esterases	

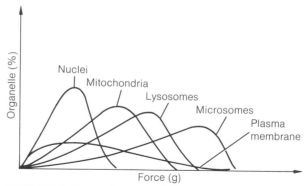

FIGURE 2.2. Organelle separation by differential centrifugation.

est centrifugal speed. With increasing centrifugal speed, the mitochondria, lysosomes, and microsomal fractions sediment in that order. Plasma membrane vesicles sediment over a wide range of centrifugal speeds due to fragmentation during homogenization. It is clear from Figure 2.2 that other techniques must be employed before highly purified subcellular fractions are obtained.

The properties of rat liver organelles are summarized in Table 2.2. The data in column 2 provide some idea of the relative proportions of the organelles in mammalian liver. For example, mitochondria represent 25% of the total cell protein, although lysosomes comprise less than 10% of this amount. It should also be noted that a mitochondrial preparation contaminated with all of the lysosomes of the cell is nevertheless more than 90% pure on a protein basis. Interestingly, the plasma membrane, which completely surrounds the cell, represents only 2% of the total protein. Although soluble proteins comprise only 30% of the total, it should be recalled that the isolated organelles contain soluble protein. Somewhat less than 50% of the total cell protein is probably membrane associated.

In column 3 of the table, the relative sizes of the organelles are indicated. It should be noted that the sedimentation behavior of an organelle in a sucrose density gradient (column 4) does *not* correlate with size but with composition. Nucleic acid is more dense than protein, and protein is more dense than lipid. These facts account for the relatively high density of nuclei and the low density of the Golgi apparatus.

Since each organelle has a specific function, it must possess a unique complement of enzymes. This prediction has been amply verified by the subcellular localization of numerous enzymes (Table 2.2, column 5). Association of a specific enzyme with an organelle greatly facilitates the assay and isolation of that organelle. It also points to the

Table 2.2   Properties of Rat Liver Organelles

Organelle	Protein (%)	Diameter (μ)	Equilibrium density in sucrose (g/ml)	Organelle-specific enzyme marker
Liver cell	100	20	1.20	—
Nuclei	15	5–10	1.32	DNA polymerase
Golgi apparatus	2	2	1.10	Glycosyl transferases
Mitochondria	25	1	1.20	Monoamine oxidase (outer membrane) Cytochrome c (inner membrane)
Lysosomes	2	0.5	1.20	Acid phosphatase
Endoplasmic reticular vesicles	20	0.1	1.15	Cytochrome b_5 reductase and cytochrome b_5; glucose 6-phosphatase
Plasma membrane	2	—	1.15	Na^+, K^+ ATPase; viral receptors
Soluble protein	30	<0.01	—	—

existence of an exceptionally complex mechanism of membrane biogenesis in eukaryotic cells.

Once the subcellular organelles have been separated, their membranes can be isolated. For those organelles enclosed by a single membrane, osmotic shock treatment followed by centrifugal separation of the membrane ghosts from the intraorganellar soluble proteins allows one to study membrane composition. Nuclei and mitochondria possess two membranes, however, and these must be separated before their chemical and physical properties can be studied. It has been found that the inner nuclear membrane can be purified after preferential disruption of the outer nuclear membrane with the neutral detergent, Triton X-100. Similarly, constituents of the outer mitochondrial membrane can be preferentially dissociated with the detergent digitonin, a steroid glycoside that inserts into and disrupts sterol-containing membranes.

Membrane Composition   With the plasma and intracytoplasmic membranes separated, biochemical analyses become possible. These analyses have shown that the two major constituents of biological membranes are proteins and lipids, some of which are derivitized by chains of sugar residues. The relative proportions of these two molecular classes vary widely depending on the source. For example, myelin membranes contain about three times as much lipid as protein, whereas bacterial membranes and the inner mitochondrial membrane are predominantly protein. Biological membranes, such as

Table 2.3   Lipid Constituents of Membranes

A.   Phospholipids

Phosphatidic acid (P.A.): $R'-\overset{\overset{\displaystyle O}{\|}}{C}-O-CH$

$$CH_2-O-\overset{\overset{\displaystyle O}{\|}}{C}-R$$

$$CH_2-O-\overset{\overset{\displaystyle O}{\|}}{\underset{\underset{\displaystyle OH}{\diagdown}}{P}}-OH^{\oplus}$$

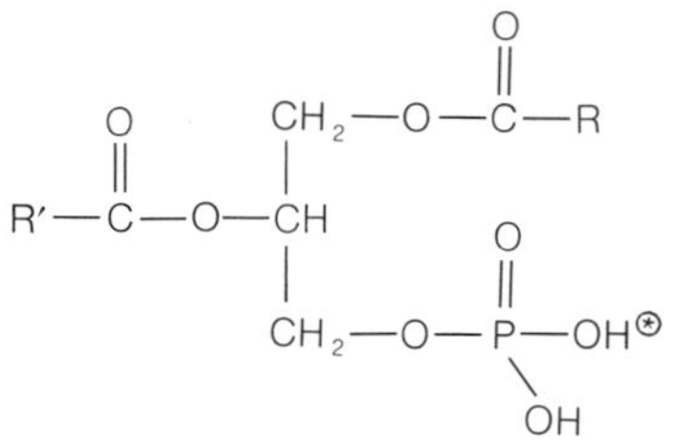

1.   Phosphatidyl serine/P.A.-serine: $-CH_2-\overset{\overset{\displaystyle NH_3^+}{\diagup}}{\underset{\underset{\displaystyle CO_2^-}{\diagdown}}{CH}}$

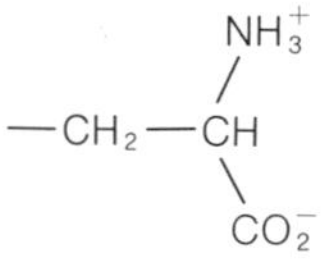

2.   Phosphatidyl ethanolamine/P.A.-ethanolamine: $-CH_2-CH_2-NH_3^+$
3.   Phosphatidyl choline/P.A.-choline: $-CH_2-CH_2-N^+(CH_3)_3$
4.   Phosphatidyl glycerol/P.A.-glycerol
5.   Cardiolipin/P.A.-glycerol-P.A.

B.   Sphingolipids

Ceramide: $R'-\overset{\overset{\displaystyle O}{\|}}{C}-NH-\overset{\displaystyle CHOH-CH=CH-R}{\underset{\underset{\displaystyle CH_2-OH^{\oplus}}{|}}{CH}}$

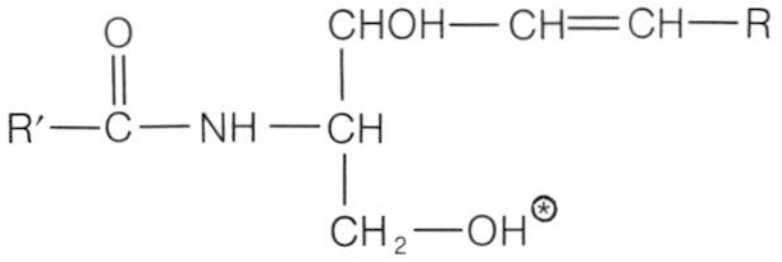

1.   Sphingomyelin: ceramide-P-choline
2.   Cerebrosides: ceramide-sugar
3.   Gangliosides: ceramide-sugars

C.   Sterols

1.   Cholesterol: $HO^{\oplus}$

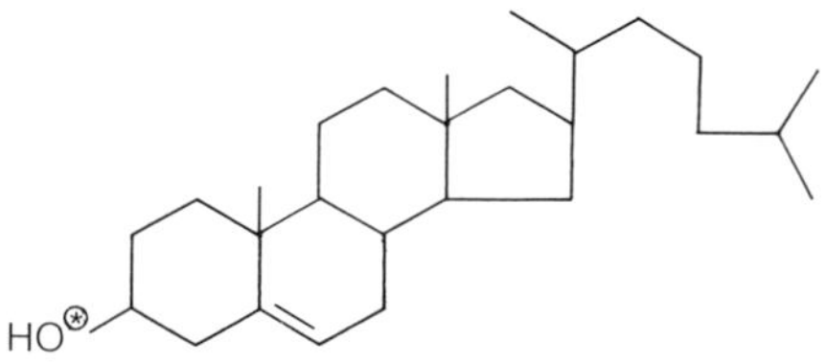

2.   Cholesterol ester

*   Site of derivitization in the complete lipid

the bacterial gas vacuole membrane, are known in which the sole constituent is a single protein.

Membrane Lipids

The principal lipid components of most biological membranes are listed in Table 2.3. Glycerophospholipids and sphingolipids comprise the polar lipids of animal cell membranes, whereas the neutral sterols comprise the principal nonpolar membrane lipids. Most phospholipids possess either anionic or zwitterionic polar head groups and have two fatty acids esterified to the glycerol moiety. One of these long chain acids usually possesses one or more *cis*-alkene double bonds, which confer upon the membrane a certain degree of fluidity (see Chapter 3). As indicated in Table 2.4, plasma membranes are far richer in cholesterol than are intracytoplasmic membranes. Eukaryotic organelles, however, appear to contain appreciable sterol, with the amount depending on the specific organelle and its source. The outer mitochondrial membrane, for example, contains more cholesterol than the inner mitochondrial membrane, and this fact provides the basis for the preferential disruption of the former by digitonin. Such neutral fats as the triglycerides, which are largely absent from biologic membranes, represent the major form of storage lipid found principally in the cytoplasm.

Although virtually all biological membranes contain a multiplicity of lipids, bacterial membranes are compositionally simpler than those of eukaryotic organisms. Phospholipids are the principal lipid constituents of the bacterial membrane, and sphingolipids and cholesterol are conspicuously absent. Moreover, the fatty acid composition of the bacterial phospholipids is relatively simple and can be altered experimentally. Since the fatty acids are, in large part, responsible for the physicochemical properties of the membrane, it is possible to study the dependence of membrane-associated, protein-catalyzed processes on the physical properties of the matrix. Such investigations have

Table 2.4 Protein and Lipid Content of Membranes

Membrane	Approximate protein: lipid ratio (wt/wt)	Approximate cholesterol: polar lipid (molar)
Myelin	0.3	1.0
Liver plasma membrane	1	0.4
Endoplasmic reticulum	1	0.06
Mitochondrial outer membrane	1	0.06
Mitochondrial inner membrane	3	0.03
Bacterial membranes	3	0.00

provided evidence on the importance of protein–lipid interactions to the normal functioning of the cell.

Membrane Proteins

The protein constituents of a membrane are of particular interest because they confer upon the structure its catalytic activities. Proteins may be loosely associated with the membrane (peripheral membrane proteins), or they may be integrated within the membrane structure (integral proteins). The former proteins can be solubilized by relatively mild treatment, and they dissociate to a lipid-free, water-soluble form. Some peripheral proteins can be solubilized by washing the membranes with solutions of high salt concentrations. For example, cytochrome c can be extracted from the inner mitochondrial membrane with 3M NaCl, and the same treatment removes acetylcholinesterase from nerve membranes. Electrostatic forces are probably important for the association of these proteins to the membrane surface. Other peripheral proteins are extracted from membranes by removal of Ca^{2+} and Mg^{2+} ions and/or inclusion of such divalent cation chelating agents as EDTA, which complexes Ca^{2+} and Mg^{2+}, or EGTA, which has preferential affinity for Ca^{2+}. Bacterial proton-translocating ATPase complexes have been extracted from membranes by this procedure. The association of these proteins may depend on divalent cation bridges between negatively charged groups of the protein and membrane. Still other peripheral proteins are released when salt is *removed* from the membrane suspension and the membranes are washed with distilled water. Possibly, in this case, electrostatic repulsive forces between integral and peripheral proteins, which are normally damped out by ionic species in the medium, are responsible for the dissociation of the membrane complexes. Alternatively, salt may be important for the maintenance of a peripheral protein conformation, which has a high affinity for a specific integral membrane protein.

Solubilization of integral membrane proteins usually requires treatment with agents that destroy the structural integrity of the membrane. Numerous *detergents* (amphipathic substances with polar and nonpolar moieties) are available for this purpose. They include negatively charged, positively charged, and neutral species. Such neutral detergents as Triton X-100, Tergitol, and Lubrol are commonly used, and they sometimes allow extraction of integral proteins in an enzymatically active form. Unfortunately, many membrane proteins remain in a particulate form after extraction with these detergents. Anionic detergents, such as sodium deoxycholate and sodium dodecyl sulfate, are more effective as membrane-solubilizing agents, but they are also more effective as protein denaturants. These agents are often employed when analyses do not require retention of enzymatic activity. Finally, cationic detergents, such as cetyl trimethyl ammonium bromide (CTAB), have been effective for the solubilization of certain membrane proteins. For example, CTAB has been used to solubilize rhodopsin from the mammalian rod outer segment membrane.

Recent studies have shed some light on the mechanisms by which detergents release integral proteins from the nonpolar phospholipid matrix. The hydrophobic portion of a *nonionic* detergent probably interacts specifically with the hydrophobic portion of the protein, substituting for the fatty acid chains of the membrane phospholipids. The protein, bound to detergent, is thereby released in an apparently molecularly dispersed form. On the other hand, an *ionic* detergent will interact with and alter the conformation of the hydrophilic as well as the hydrophobic moiety of the protein, and, since the former part of the protein is usually responsible for catalytic activity, biological activity is lost. This mechanism is in agreement with the observation that only ionic detergents bind to and denature simple soluble proteins. Little interaction is observed between typical hydrophilic proteins and a nonionic detergent.

The second group of agents used to solubilize biological membranes are the aqueous denaturants or *chaotropes*, which include urea, guanidinium chloride, sodium iodide, and sodium thiocyanate. They function by introducing "chaos" into biological membranes, in part by disrupting the ordered arrangement of water molecules that surround these structures. The Na^+, K^+-translocating ATPase from kidney membranes has been solubilized by the use of chaotropes, and the mitochondrial electron transport chain has been extensively fractionated.

Several other agents are available for membrane protein solubilization. *Organic solvents* have been used both separately and in conjunction with detergents and chaotropes. Alcohols of intermediate chain length, such as butanol, have been used to isolate enzymatically active proteins, whereas basic or acidic solvent systems have been used to solubilize and denature membrane proteins. Protein modification, which follows treatment of membranes with chemical agents or proteolytic enzymes, may also facilitate solubilization, but these procedures invariably cause partial or total loss of biological activity.

Once integral membrane proteins have been solubilized, fractionation by conventional biochemical techniques is possible. When continued solubilization requires the presence of a detergent, chaotrope, or organic solvent, the purification procedure must be carried out in the presence of these agents. Ammonium sulfate and isoelectric precipitation, gel filtration, ion exchange chromatography, and preparative electrofocusing and gel electrophoresis have all been applied to integral membrane proteins in the presence of suitable solubilizing agents.

The molecular weights of membrane proteins and the complexity of a specific membrane can be estimated by the technique of polyacrylamide gel electrophoresis after complete solubilization of the membrane in sodium dodecyl sulfate has been achieved. Vigorous treatment of a membrane with this anionic detergent solubilizes and denatures the constituent proteins. The hydrophobic side chains of the detergent interact with the nonpolar portions of the protein, so that the macromolecule becomes coated with negatively charged molecules. Electrostatic repulsive forces cause the protein to assume a conformation that is maximally extended, and it takes the shape of a long thin cylinder. Since the number of detergent molecules, and, hence, the number of negative charges, is dependent only on the size of the protein, the molecule should migrate in an electric field on the basis of its size alone. That this is the case has been verified for proteins of molecular weights between 15,000 and 150,000, although strongly basic proteins and glycoproteins with large carbohydrate content may behave anomalously. Within this size range, the mobility of a normal sodium dodecyl sulfate-coated protein in a polyacrylamide gel is approximately proportional to the inverse of the logarithm of the molecular weight. This technique has been extensively used for membrane protein analyses.

With this procedure, it has been possible to show that myelin and mammalian muscle sarcoplasmic reticular membranes each contain three major proteins. The most abundant sarcoplasmic reticular protein has been identified as the calcium-translocating ATPase, whereas a single protein, rhodopsin, is the major species of the rod outer segment membrane of the mammalian retina. By contrast, bacterial cytoplasmic membranes are far more complex, containing more than 100 different protein species with none predominating. The large number of bacterial membrane proteins was predicted from the multiplicity of membrane-related functions. A generalization that seems to emerge from these observations is that cellular differentiation and specialization render a membrane structurally less complex.

A few integral membrane proteins have been purified and studied. Liver microsomes contain an electron transport chain, which appears to function in the desaturation of fatty acids. Electrons pass sequentially from reduced nicotinamide adenine dinucleotide (NADH) to cytochrome b_5 reductase, then to cytochrome b_5, and, finally, to the desaturase enzyme. Both the reductase and cytochrome b_5 have been purified to homogeneity and structurally characterized. The former is a flavin enzyme with a molecular weight of 43,000. The latter is a heme protein with a molecular weight of 16,700 (Figure 2.3). Mild trypsin treatment splits each of these proteins into two parts: a water-soluble moiety possessing full enzymatic activity and an inactive, water-insoluble peptide. In each case, the water-insoluble moiety is the smaller of the two. The amino acid compositions of these polypeptide chains are summarized and compared with the amino acid composition of bulk membrane protein in Table 2.5. It can be seen that the hydrophilic portions are rich in basic and acidic amino acid residues, whereas the hydrophobic regions are greatly enriched for the non-polar amino acid residues. This observation is particularly striking for the small cytochrome b_5 molecule. It seems unlikely, however, that composition alone accounts for the properties of these molecules. Possibly, most of the polar amino acid side chains of the hydrophilic moieties point toward the periphery of the protein, whereas the nonpolar side chains are directed toward the center of the globular structure. The reverse may be true of the hydrophobic portions: The hydrophobic amino acid residues may be in association with membrane lipids or other membrane proteins, whereas the hydrophilic residues may be exposed to the aqueous environment or buried internally. In view of the structural features noted above for these proteins, the

FIGURE 2.3. Fractionation and properties of the protein components of the liver microsomal electron transport chain.

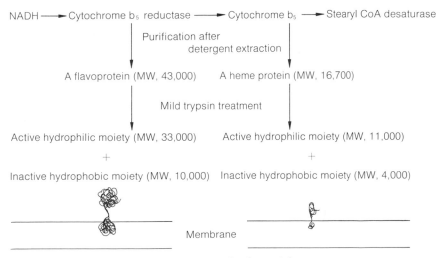

NADH ⟶ Cytochrome b_5 reductase ⟶ Cytochrome b_5 ⟶ Stearyl CoA desaturase

Purification after detergent extraction

A flavoprotein (MW, 43,000)   A heme protein (MW, 16,700)

Mild trypsin treatment

Active hydrophilic moiety (MW, 33,000)   Active hydrophilic moiety (MW, 11,000)

+   +

Inactive hydrophobic moiety (MW, 10,000)   Inactive hydrophobic moiety (MW, 4,000)

Membrane

Presumed structure of native protein

Table 2.5 Amino Acid Composition of Membrane Protein Fractions (%)

Membane fraction	Basic: Lys, Arg, His	Acidic: Asp, Glu	Semipolar: Thr, Ser, Pro, Gly, Ala, Cys	Hydrophobic: Val, Met, Ile, Leu, Tyr, Phe, Trp
Total membrane (red blood cell, plasma, endoplasmic reticulum)	12–14	19–21	32–35	32–35
Peripheral proteins	14	25	30	30
Integral proteins	10	15	36	39
Cytochrome b₅ (hydrophilic moiety)	20	25	30	25
Cytochrome b₅ (hydrophobic moiety)	5	12	33	50
Cytochrome b₅ reductase (hydrophilic moiety)	16	19	33	32
Cytochrome b₅ reductase (hydrophobic moiety)	12	15	26	46

function of each protein moiety appears clear. The hydrophilic moiety confers upon the molecule its catalytic activity, whereas the hydrophobic moiety serves as an anchor to maintain proper topography on the membrane surface.

Another well-characterized membrane protein is glycophorin, the major glycoprotein in the red blood cell membrane. This protein can be isolated from this membrane after treatment with chaotropes. It has a molecular weight of 55,000 and is 40% protein, 60% carbohydrate on a weight basis. It bears the ABO- and MN-blood group antigenic specificities of the cell and serves as the receptor for influenza virus. The probable structure of this glycoprotein is shown in Figure 2.4. It consists of a single polypeptide chain to which short carbohydrate chains are covalently attached to amino acid residues comprising the N-terminal region of the protein. This portion of the molecule is polar by virtue of the presence of sugar and hydrophilic amino acid residues. The carboxyl end of the polypeptide is also rich in polar amino acid residues, but the central region of the protein contains an extremely hydrophobic stretch of 25 amino acid residues. Radioactive labeling experiments have shown that the carbohydrate-rich N-terminus is localized on the external surface of the red blood cells, whereas the carboxyl terminus is exposed to the cytoplasm. These observations suggest that the molecule penetrates the membrane.

Recent studies have shown that the hydrophobic portion of the molecule has a high affinity for phospholipids

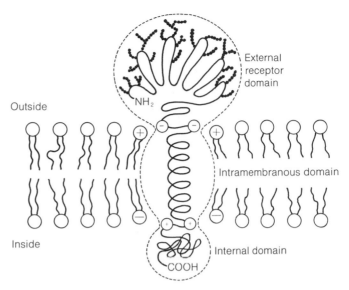

Outside

Inside

NH₂

External receptor domain

Intramembranous domain

Internal domain

COOH

FIGURE 2.4. Presumed structure of glycophorin in the mammalian red blood cell plasma membrane.

After Segrest and Kohn (1973).

and cholesterol, the two principal lipid constituents of the red blood cell membrane. Moreover, this portion of the molecule forms a very stable α helix in a hydrophobic environment. The length of this α helix is about 40 Å, somewhat less than the known width of biologic membranes. These observations provide strong experimental evidence for the structural model proposed in Figure 2.4.

Other proteins have been examined from topologic and structural standpoints. Thus, current evidence on the topography of the electron transport chain in the inner mitochondrial membrane indicates that some integral proteins are localized at the inner face, some at the outer face, others may be completely embedded within the membrane matrix, and still other proteins appear to penetrate the entire structure. Peripheral membrane proteins, however, are probably associated primarily with complementary portions of integral proteins, which are exposed to one of the two surfaces of the bilayer. As the structures and topographies of increasing numbers of membrane proteins become known, insight concerning the relationships between membrane biogenesis, structure, and function should be forthcoming.

Selected References

Fleisher, S. and L. Packer (eds.). "Isolation of purified subcellular fractions and derived membranes," in *Methods in Enzymology XXXI, XXXII.* Academic Press, New York, 1974.

Law, J. H. and W. R. Snyder. "Membrane lipids," in *Membrane Molecular Biology* (C. F. Fox and A. D. Keith, eds.). Sinauer Associates, Inc., Stamford, Conn., 1972, p. 3.

Machtiger, N. A. and C. F. Fox. "Biochemistry of bacterial membranes," in *Annual Review of Biochemistry, Vol. 42*. Annual Reviews, Inc., Palo Alto, Ca., 1973, p. 575.

Marchesi, V. T. "Molecular orientation of proteins in membranes," in *Biochemistry of Cell Walls and Membranes* (C. F. Fox, ed.), (MTT, International Review of Science, Biochemistry Series, Vol. 2). Medical Technical Publishers, Ltd., Aglesbury, England, 1975, p. 123.

Osborn, M. J., J. E. Gander, E. Parisi, and J. Carson. Mechanism of assembly of the outer membrane of *Salmonella typhimurium.* Isolation and characterization of the cytoplasmic and outer membranes. *J. Biol. Chem., 247*:3962 (1972).

Racker, E. "The two faces of the inner mitochondrial membrane," in *Essays in Biochemistry, Vol 6*. Academic Press, London, 1970, p. 1.

Rogers, M. J. and P. Strittmatter. Evidence for random distribution and translational movement of cytochrome b_5 in endoplasmic reticulum. *J. Biol. Chem., 249*:895 (1974).

Segrest, J. P. and L. D. Kohn. "Protein-lipid interactions of the membrane penetrating MN-glycoprotein from the human erythrocyte," in *Protides of the Biological Fluids—21st Colloquim* (H. Peeters, ed.). Pergamon Press, New York, 1973, p. 183.

Spatz, L. and P. Strittmatter. A form of NADH—Cytochrome b_5 reductase containing both the catalytic site and an additional hydrophobic membrane-binding segment. *J. Biol. Chem., 248*: 793 (1973).

Steck, T. L. "Membrane isolation," in *Membrane Molecular Biology* (C. F. Fox and A. D. Keith, eds.). Sinauer Associates, Inc., Stamford, Conn., 1972, p. 176.

Steck, T. L. The organization of proteins in the human red blood cell membrane. *J. Cell Biol., 62*:1 (1974).

Steck, T. L. and C. F. Fox. "Membrane proteins," in *Membrane Molecular Biology* (C. F. Fox and A. D. Keith, eds.). Sinauer Associates, Inc., Stamford, Conn., 1972, p. 27.

3 Structure of Membranes and Serum Lipoprotein Complexes

For is and is-not come together;
Hard and easy are complementary;
Long and short are relative;
High and low are comparative;
Pitch and sound make harmony;
Before and after are a sequence.

Lao Tzu

In Chapter 2, the principal constituents of biological membranes, lipids and proteins, were examined from a structural standpoint. It was noted that membrane components are, in general, amphipathic; they possess both polar and nonpolar moieties. Thermodynamic considerations lead to the conclusion that such macromolecules in an aqueous environment should associate spontaneously with the formation of structures of maximal stability in which nonpolar elements comprise one phase, polar elements, the other. Since the hydrophobic environment resembles an oil, the structure should be fluid—dynamic—constantly in motion. In this chapter, these predictions are shown to be valid for many biologic membranes. An appreciation of the structural aspects considered below will prove to be one of the keys to the understanding of the intricacies of membrane functions.

Soap Molecules in Aqueous Solution

Soap molecules are the alkali metal salts of long chain fatty acids. Due to the hydrophobic nature of the alkane tail and the hydrophilic nature of the negatively charged carboxyl group, these molecules exhibit "schizophrenic" behavior: part of the molecule wishes to enter a polar environment while the other moiety prefers a nonpolar surrounding. The

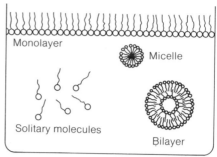

FIGURE 3.1

Stable structures formed in aqueous soap
solutions.

molecule is said to be amphipathic. When a small number
of soap molecules are dissolved in an aqueous solution,
they enter the aqueous phase by virtue of their polar carboxyl groups. Their solubility, however, is low due to the
properties of the hydrophobic tails. When more soap is
added to the solution, the fatty anions distribute at the air–
water interface, forming a *monolayer* in which the hydrophilic carboxylate anions enter the aqueous phase, while
the hydrophobic tails point upward into the nonpolar
gaseous phase. As increasing numbers of soap molecules
are added to the solution, the entire surface is covered with
a monolayer. At this point, both the solution and the air–
water interface are saturated with fatty carboxylate anions,
yet more soap can be added. The structures that result are
spherical *micelles* in which the carboxyl groups of the soap
molecules face the aqueous environment, while the long
chains extend inward toward the center of the micelle (Figure 3.1). The alkane chains comprise a nonpolar environment and render the micellar structure thermodynamically
stable.

If an aqueous suspension of soap micelles is agitated,
soap bubbles with water inside and out result. The soap
molecules in these bubbles are arranged in a simple *bilayer*
(Figure 3.1). As for the micelle, the carboxyl groups extend
into the aqueous phase, while the side chains comprise a
hydrophobic environment. Since soap micelles and bilayers
form spontaneously and are stable in aqueous solution, it
is possible to conclude that these are among the most
stable structures soap molecules can form.

**Biological
Membranes**

In the previous chapter, it was noted that, like soap molecules, the principal constituents of biological membranes,
polar lipids and proteins, have amphipathic properties. In
view of these properties, it is not surprising that bilayers
and micelles have important counterparts in biology. In-
deed, they represent the fundamental structural elements of
biologic membranes and lipoprotein complexes, respectively.

The current conception of biological membranes is shown in Figure 3.2. The proposed model suggests that the essential structural repeating unit is the phospholipid molecule in a bilayer arrangement. Globular proteins may be inserted into the bilayer in a seemingly random fashion. Some proteins (such as cytochrome b_5) are localized at one or the other of the two surfaces of the lipid bilayer; other proteins (such as glycophorin) may pass from one side of the membrane to the other; and still other proteins may be embedded in the hydrophobic matrix. Although most of the membrane phospholipids are in the bilayer array, a considerable number may be closely associated with integral membrane proteins. Evidence for this model comes from the observation that the stability and activity of many membrane enzymes depend on lipid association.

Superimposed upon this mosaic structure, a variety of peripheral proteins may be loosely associated with the membrane as a result of interactions with integral protein or lipid constituents. In fact, the interrelationships between membrane proteins and the cell skeletal structure, the microtubules and microfilaments, is an active area of investigation. Finally, the entire structure is thought to be dynamic, rather than static, with most components capable of lateral and rotational diffusion. Thus, Figure 3.2 depicts the fluid mosaic model of a biological membrane at one point in space and time.

FIGURE 3.2. The lipid–globular protein mosaic model of a biological membrane with a lipid matrix.

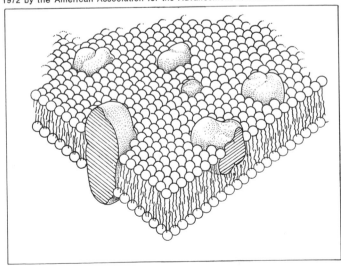

**Serum
Lipoprotein
Complexes**

Just as the bilayer provides biological membranes with an element of structure, serum lipoprotein complexes are structurally related to simple micelles. Lipoprotein complexes represent the vehicles for the plasma transport of phospholipids, triglycerides, cholesterol, and cholesterol esters. Although these complexes fall into several categories, depending on their lipid and protein composition, we will be concerned with only one such micellar structure, that of the high-density lipoprotein complex, which is depicted in Figure 3.3. The principal phospholipids, phosphatidyl choline and sphingomyelin, are arranged with the polar head groups exposed to the medium with hydrophobic side chains projecting inward toward the center of the micelle. Cholesterol molecules may be sandwiched between the phospholipids. In the center of the structure exists a neutral core, which consists primarily of cholesterol esters and triglycerides, the most nonpolar elements of the complex. Hydrophobic forces are thought to hold the entire structure together.

Several proteins are associated with the micelle. Among these is a protein known as apolipoprotein AII. Like cytochrome b_5, this protein has many hydrophobic amino acid residues, but these residues are interspersed throughout the linear sequence. The protein can be prepared without lipid, and when it is recombined with phospholipid, its helical content increases. This behavior is reminiscent of that of glycophorin, as discussed in Chapter 2. When the α helix is formed, one finds, from sequence analyses, that most of the hydrophobic residues are on one side of the helix, while most of the charged and strongly polar amino

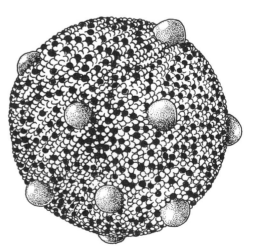

FIGURE 3.3

Proposed model of the serum high density
lipoprotein complex. Sphingomyelin (●);
phosphatidyl choline (○).

After Assmann and Brewer (1974).

acid residues are on the other side (Figure 3.4). Very possibly, the nonpolar side of the helix is in contact with lipid molecules in the intact complex. Throughout the length of the polypeptide chain, one finds prolyl residues occurring at approximately 20 amino acid residue intervals. Since proline residues effectively break α helices, it is suggested that the protein exists as a series of short helices, each segment having a length of about 20 A. Interestingly, this is the approximate length of an extended 18-carbon fatty acid. Possibly the two classes of macromolecules, protein and phospholipid, associate along their long axes.

Apolipoprotein AII represents an integral protein of the micelle and associated with it is a peripheral protein, the apolipoprotein AI. The latter protein apparently binds specifically to apolipoprotein AII and not to lipid. Curiously, this peripheral protein has been shown to influence a spe-

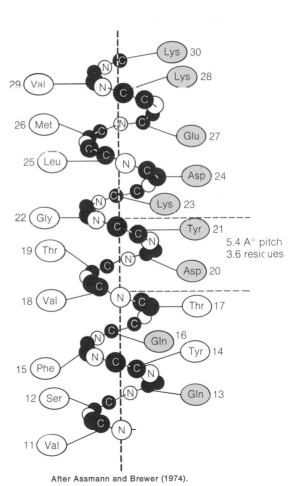

FIGURE 3.4

Structure of a portion of the polypeptide chain of apolipoprotein A II arranged in an α-helix.

5.4 A° pitch
3.6 residues

After Assmann and Brewer (1974).

cific enzymatic activity. It activates the lipid metabolic enzyme, phosphatidyl choline:cholesterol acyltransferase.

Other proteins are known to interact specifically with the high density lipoprotein complex. For example, C-type proteins are easily dissociated from the complex. Whereas apolipoprotein CIII appears to interact directly with phospholipids, apolipoprotein CII, which activates phospholipid lipase, is bound to the complex through apolipoprotein CIII. The picture that emerges is that of a fluid mosaic micelle in which integral and peripheral globular proteins are inserted in a micellar matrix of lipid. A principal function of the integral proteins may be to bind specific functional peripheral proteins to the structure and, perhaps, to modulate their catalytic activities.

Fluidity of Membrane Constituents

The fluid nature of many biological membranes and lipoprotein complexes has been established by a variety of physical techniques. For example, application of X-ray diffraction and differential scanning calorimetry to bacterial membranes of relatively simple lipid composition has shown that these structures can pass through a thermal phase transition. At a temperature below the transition, the fatty acid side chains of the phospholipids are in a relatively rigid crystalline state; at a temperature above the transition, these side chains assume a far more fluid, random structure (Figure 3.5). At temperatures in between, both fluid and solid lipid phases exist. Although, in the technique of X-ray diffraction, the degree of ordering of the fatty acid side chains in the membrane is measured, differential scanning calorimetry measures the energy necessary to increase the temperature of a membrane suspension. Just as energy is required to melt ice, energy must be absorbed by a biological membrane to "melt" the crystalline array of its fatty acid side chains. Calorimetry measures energy absorption as a function of temperature. Application of these techniques to artificial membranes has shown that the temperature range of the phase transition is determined largely by the fatty acid composition of the membrane. The chain length, the degree and position of unsaturation, and the configuration of the unsaturation (*cis* or *trans*) are important factors. Longer chain lengths lead to greater stability of the crystalline state and, hence, to higher transition temperatures,

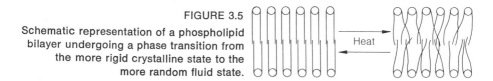

FIGURE 3.5

Schematic representation of a phospholipid bilayer undergoing a phase transition from the more rigid crystalline state to the more random fluid state.

whereas any discontinuity in the regular array of fatty acid side chains (i.e., unsaturation or a branched fatty acid chain) leads to instability and, therefore, to a lower transition temperature.

Studies on membranes isolated from living cells have also led us to interesting conclusions. First, the temperature range over which a phase transition occurs is determined largely by the membrane fatty acyl groups present and the ratio of saturated to unsaturated fatty acids. Second, the temperature range is ordinarily very broad because of the heterogeneity of the fatty acyl groups present in the phospholipids. Phase changes of biological membranes occur over a minimal 10° range and may be so broad that a distinct transition cannot be measured. Third, most, if not all, living organisms maintain a lipid composition that renders the membrane semifluid at the growth temperature. This appears to be necessary because such processes as membrane biogenesis, transmembrane transport, and exo- and endocytosis can only occur when the membrane exhibits sufficient fluidity. But, if a membrane is too fluid, it may lack the necessary integrity. As a consequence, living organisms have evolved intricate mechanisms to regulate the degree of unsaturation in their phospholipids. These mechanisms ensure a proper degree of membrane fluidity for the normal functioning of essential membrane-associated cell processes.

Detailed quantitative information concerning artificial and biological membranes has resulted from the application of spectroscopic techniques. An appropriate probe (Figure 3.6) may be inserted into the membrane of interest. A fluorescent probe, for example, absorbs energy at one wavelength and emits light of lower energy. Analyses of the polarization of the emitted light provide information about the orientation and the rotational and translational mobility

FIGURE 3.6 Spectroscopic probes used for membrane analyses. (A) 10-anthranyl palmitate, a fluorescent probe; (B) 6-nitroxide palmitate, an ESR probe.

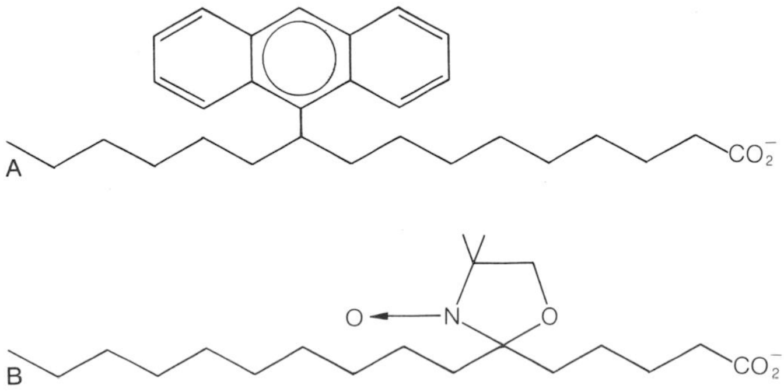

of the probe. Orientation and mobility of the probe, in turn, reflect the nature of the membrane environment in which the probe has been embedded.

The technique of electron spin resonance (ESR) involves measurement of the energy absorbed by a free spinning electron. Since biological membranes normally lack molecules with unpaired electrons, the application of this technique requires the introduction into the membrane of a stable molecular grouping with a free electron. Nitroxides are frequently used for this purpose (Figure 3.6). The unpaired electron possesses a magnetic moment, which tends to align in a magnetic field in the same way a magnet does. From quantum theory we know that the electron can assume two possible energy states that depend on the local magnetic field. Energy absorption causes the electron to "resonate" between these two states. The energy absorbed is influenced by the atomic nucleus in the immediate vicinity of the absorbing electron (the nitrogen nucleus) and the environment in which the probe finds itself. The nitrogen nucleus of the nitroxide splits the absorption maximum into three peaks, whereas environmental factors influence the broadness, shape, and relative positions of these peaks. If all probes can rotate freely and are in a similar environment, three sharp absorption peaks will result. Thus, electron spin resonance profiles, as fluorescence measurements, provide information on membrane orientation and fluidity.

Spectroscopic techniques have been used to quantitate membrane fluidity. For example, it has been shown that a phospholipid membrane exhibits greater fluidity in the center of the bilayer than near its periphery, where the polar head groups maintain a certain degree of rigidity. Lateral diffusion of a phospholipid in a typical membrane at body temperature is characterized by a diffusion constant of about 10^{-7} cm^2/s, suggesting that lipids can diffuse from one side of an animal cell to the other in a few minutes. The rate at which a phospholipid can flip from one side of a bilayer to the other, however, is very low. The half-time for the flipping of a phosphatidyl choline molecule in a phospholipid vesicle has been estimated to be about 20 h at 30°C. Other measurements have led to the suggestion that about one-half of the surface of a bacterial membrane is coated with protein, and that roughly one-fifth of the membrane phospholipid is associated with integral proteins. These same techniques have been useful for demonstrating the biologically important effects of anesthetics and tranquilizers on membrane lipid fluidity. It is anticipated that further application of these procedures will provide valuable information concerning the fine structure of biological membranes.

Although the application of physical techniques to membranes has clearly established the fluid nature of the lipid constituents, other approaches have shown that membrane proteins may also be in a fluid state and be capable of rapid lateral diffusion and rotation about the axis perpendicular to the plane of the membrane. Mouse and human cells were fused in an experiment in which an inactivated envelope virus was employed to induce the fusion event. The diffusion of surface glycoproteins was then followed as a function of time, employing fluorescent antibodies directed against either the mouse or the human cell surface components. Immediately following fusion, the mouse antigens were localized to one hemisphere of the fused cell, and the human antigens were restricted to the other. After 30 min at 37°C, however, both mouse and human antigens had spread over the entire surface of the cell. Since antigen movement appeared to be due to simple lateral diffusion rather than removal from and insertion into the membrane, it appeared that both large and small integral membrane components were capable of translational motion. Thus, most biological membranes are perhaps best pictured as dynamic structures in which components are in continual motion. Membrane proteins will be held in a rigid state only if highly specific associations between them can occur.

Membrane Action of Anesthetics

Many biological membranes are thought to contain ion-specific "channel" proteins, which render the structure selectively permeable to particular ionic species. As a result of selective ionic permeability and an unequal distribution of ions across the phospholipid barrier, a transmembrane electrical potential, the *resting membrane potential*, is generated (Chapters 5–7). A sudden increase in the permeability of the nerve or muscle cell membrane to Na^+, corresponding to the opening of the Na^+ channel, may initiate an *action potential* a "spike" of depolarization (Chapter 7), and the opening of the Na^+ gate may well be due to a protein conformational change, which is blocked by drugs known as *anesthetics*. For the purposes of our discussion, an anesthetic can be defined as a compound that, at an appropriate concentration, reversibly blocks an action potential without altering the resting membrane potential.

Since the nature of the ionic channels that determine the action potential is not well understood, it is impossible to define in precise molecular terms how anesthetics exert their effects. Nevertheless, several studies have led to the conclusion that anesthetics act within the membrane of a nerve or muscle cell. In fact, the effectiveness of a compound in blocking an action potential is related to its solubility in a hydrophobic environment. This observation has led to the famous *Meyer–Overton rule of anesthesia*, which states that narcosis occurs when any chemically indifferent substance attains a certain molar concentration in the membrane. This rule may be an approximation to the true situation, since the size of the anesthetic molecule appears to be of some importance. That anesthetics act primarily through the medium of the membrane is further suggested by the observation that these drugs protect red blood cell membranes from hypotonic lysis and stabilize the membranes of lysosomes and catecholamine-containing vesicles.

Application of physical techniques has revealed that anesthetics expand the membrane and increase its fluidity. Although the precise mechanism is not yet established, it would appear reasonable that insertion of foreign molecules into the phospholipid bilayer would introduce disorder, disrupting the regular array of phospholipid molecules. A more fluid hydrophobic matrix might secondarily induce conformational changes in the integral membrane proteins. In other words, anesthetics may act on membrane proteins through the medium of the phospholipid matrix.

Anesthetics, as we have defined them, comprise a diverse collection of compounds. Typical anesthetics, tranquilizers, and sedatives have a general quieting influence. Such drugs as the anticonvulsants, vasodilators, and antihistamines, however, exert relatively specific effects on the central or peripheral nervous system, whereas such "social drugs" as ethanol, narcotics, detergents, and a variety of environmental pollutants sometimes have quite bizarre effects. The clinical specificities of these drugs are explained only partially by the general properties described above. Some anesthetics are apparently preferentially concentrated by specific cells of the nervous system and, therefore, exert local effects. Another anesthetic may interact specifically with a membrane-associated enzyme or receptor molecule to influence its activity. Additionally, an anesthetic may be neutral or bear a positive or negative charge. The charge on the molecule can influence, for example, the degree to which Ca^{2+} is complexed to the membrane, and Ca^{2+} is known to influence a variety of physiologic processes. Specificity may also result from an asymmetric distribution of the anesthetic molecules within the phospholipid bilayer. Thermodynamic considerations lead to the conclusion that an amphipathic drug will intercalate preferentially with its hydrophobic moiety imbedded in the membrane and its hydrophilic group exposed to the aqueous solution at one of the surfaces of the bilayer. Since phospholipids are distributed asymmetrically across the red blood cell membrane, with the negatively charged phosphatidyl serine molecules

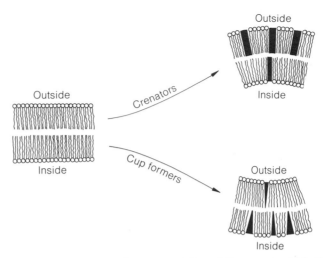

FIGURE 3.7. Schematic representation of the proposed binding of amphipathic compounds that are crenators or cup formers to the phospholipid regions of the red blood cell membrane. Crenators intercalate preferentially into the exterior half of the bilayer, causing it to expand relative to the cytoplasmic half, thereby producing crenation. Cup formers do the reverse.

After Sheetz and Singer (1974).

localized primarily in the monolayer facing the cytoplasm, one would expect that if a negatively charged drug could cross the membrane freely, it would intercalate preferentially in the outer monolayer, due to electrostatic repulsion. Conversely, since the outer monolayer contains primarily phosphatidyl choline and ethanolamine, positively charged molecules would predominate in the inner monolayer, due to electrostatic attraction. An excess of foreign molecules in the outer monolayer would tend to expand this layer relative to the cytoplasmic half and thereby *crenate* the surface of the membrane. The reciprocal effect would be expected for positively charged anesthetics, which should be *cup formers*. These predictions, which are illustrated in Figure 3.7, have been amply substantiated by recent experimentation. Anesthetics may be useful not only for clinical purposes, but also for revealing important details of membrane structure and function.

Selected References

Assmann, G. and H. B. Brewer, Jr. A molecular model of high density lipoproteins, *Proc. Nat. Acad. Sci. US*, 71:1534 (1974).

Edidin, M. Rotational and translational diffusion in membranes, *Ann. Rev. Biophys. Bioeng.*, 3:179 (1974).

Fox, C. F. and A. D. Keith (eds.). *Membrane Molecular Biology*. Sinauer Associates, Inc., Stamford, Conn. 1972.

Jackson, R. L., J. D. Morrisett, and A. M. Grotto, Jr. "Lipid-protein interactions in human plasma high density lipoproteins," in *Protides of the Biological Fluids, 21st Coll. Brugge* (H. Peeters, ed.). Pergamon Press, Oxford, 1973.

Seeman, P. The membrane actions of anesthetics and tranquilizers, *Pharmacol. Rev. 24*:583 (1972).

Sheetz, M. and S. J. Singer. Biological membranes as bilayer couples. A molecular mechanism of drug–erythrocyte interactions, *Proc. Nat. Acad. Sci. US, 71*:4457 (1974).

Singer, S. J. "Molecular biology of cellular membranes with applications to immunology," in *Advances in Immunology, Vol. 19.* Academic Press, Inc., New York, 1974, p. 1.

Singer, S. J. "The molecular organization of membranes," in *Annual Review of Biochemistry, Vol. 43.* Annual Reviews, Inc., Palo Alto, Ca. 1974, p. 805.

Singer, S. J. and G. L. Nicolson. The fluid mosaic model of the structure of cell membranes, *Science, 175*:720 (1972).

4 Biological Consequences of Membrane Fluidity and Fusion

There is a willow grows aslant a brook,
That shows his hoar leaves in the glassy stream;
There with fantastic garlands did she come
Of crew-flowers, nettles, daisies, and long purples,
That liberal shepherds give a grosser name,
But our cold maids do dead men's fingers call them:
There, on the pendent boughs her coronet weeds
Clambering to hang, an envious sliver broke;
When down her weedy trophies and herself
Fell in the weeping brook.

William Shakespeare

In the previous chapter, we saw that the lipid constituents of membranes confer on these structures a fluid character. Membrane fluidity allows for a variety of biological functions. These include cell infection by envelope viruses; cell fusion, as it occurs in the biogenesis of a muscle fiber (a myotube); and the formation of junctions between cells, to allow intercellular communication. It is also clear that exo- and endocytosis would be impossible in cells possessing rigid membranes.

Moreover, individual membrane constituents (proteins and phospholipids) appear to be synthesized and degraded independently of other membrane components, which implies a dynamic mechanism of insertion and removal, presumably dependent on fluidity. In this chapter, these processes are discussed, and mechanisms consistent with known membrane structural features are proposed.

Membrane
Biogenesis

Membrane biogenesis is one of the more controversial areas of membrane biology. The process has been studied with both mammalian and bacterial systems, and extensive work has provided information on mitochondrial and chloroplast biogenesis. This discussion will deal primarily with the biogenic mechanism in bacteria, although the principles that have emerged are likely to prove applicable to all biologic systems.

Early work on the synthesis of rat liver microsomal and plasma membranes established that the constituents of these structures are not synthesized coordinately. For example, levels of glucose 6-phosphatase are enhanced about 30-fold in rat liver microsomal membranes within five days of birth. During the same period, NADH:cytochrome c reductase and cytochrome b_5 are also induced, but the induction of these two electron carriers is neither coordinate nor equivalent in degree. Furthermore, some membrane proteins, but not others, are synthesized in increased amounts after treatment with phenobarbital.

Membrane constituents also appear to be degraded independently of one another. For example, phospholipids have a shorter half-life than proteins by a factor of two or three. Furthermore, the turnover rates of membrane proteins can be correlated with their molecular sizes, there being an inverse relationship between size and half-life. Consequently, it has been concluded that membranes are not synthesized as a unit; rather, synthesis, insertion, removal, and degradation of components may occur continuously in a dynamic fashion. Since considerable evidence suggests that the membrane itself is the site of synthesis and degradation of the membrane constituents, insertion or removal of intact constituents from membranes may never occur as such.

Although these conclusions appear applicable to bacterial as well as to animal membranes, the application of genetic techniques to bacteria has allowed far more extensive analyses of membrane biogenesis in these organisms. Bacterial mutant strains that can neither synthesize nor degrade glycerol (glycerol auxotrophs) are dependent on exogenous glycerol for the synthesis of glycerophospholipids. Upon removal of glycerol from the medium of such a mutant, net phospholipid synthesis ceases, but protein synthesis may continue for about one generation. Conversely, if the protein synthesis inhibitor, chloramphenicol, is added to the bacterial suspension in the presence of glycerol, net membrane protein synthesis is blocked, but phospholipids continue to be synthesized and incorporated into membranes for about one generation. In either case, a membrane with an altered protein to lipid ratio results.

These observations confirm the conclusion, noted above for liver microsomal membranes: protein and phospholipid syntheses are not obligatorily coupled.

In an extension of these studies, it was shown that a protein that functions in transport could be inserted into a membrane in the absence of net lipid synthesis, but that a lipid deficiency might render the protein nonfunctional. The organism used for these studies was *Staphylococcus aureus*; the transport system studied was the lactose permease. Lactose is transported into this bacterium via the lactose phosphotransferase system, which catalyzes both the transport and phosphorylation of lactose (see Chapter 5 for details). The membrane protein that catalyzes this reaction can be assayed for transport activity *in vivo* and for phosphorylation activity *in vitro*:

To facilitate interpretation of the results, a glycerol auxotroph that could synthesize the lactose phosphotransferase in the *absence of inducer* at 44°, but not at 34°C, was used. Cells were grown at 34°C in the presence of glycerol, with the result that synthesis of membranes that lacked the lactose phosphotransferase occurred. Subsequently, the cells were washed free of glycerol, transferred to fresh medium at 44°C, and allowed to grow either in the presence or absence of glycerol. In the presence of glycerol, the induction of transport activity followed phosphorylation activity. In the absence of glycerol (and, therefore, in the absence of net phospholipid synthesis), transport activity lagged behind phosphorylation activity (Figure 4.1). If, subsequently, glycerol was added back to the lipid-deficient cells in the presence of chloramphenicol, phospholipid synthesis was initiated in the absence of protein synthesis, and transport activity was partially restored (Figure 4.1). These results suggested that, although insertion of the lactose permease

FIGURE 4.1

Insertion of the lactose phosphotransferase into the membrane of S. *aureus* in the presence and absence of net phospholipid synthesis. The solid line depicts *in vitro* phosphorylation of lactose. The dashed line indicates lactose transport activity *in vivo*.

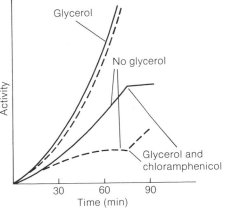

protein in *S. aureus* was not dependent on net phospho-lipid synthesis, transport activity was markedly inhibited by phospholipid depletion. Possibly, a minimal phospholipid content sufficient to guarantee membrane fluidity was nec-essary for efficient transport to occur.

If membranes are truly fluid structures, newly synthe-sized proteins and lipids should rapidly undergo lateral diffusion and randomize. Whether or not this actually oc-curs in bacterial membranes is still controversial. But, data from several lines of experimentation indicate that newly synthesized protein experiences the environment of the lipid synthesized before the insertion of that protein. That is, the protein may be inserted into preformed membrane. Either randomization of phospholipids appears to occur after their insertion, or phospholipids are inserted randomly.

Membrane fluidity may be necessary both for insertion and maximal stability of the lactose permease protein in *Escherichia coli* as suggested by studies with a mutant that could not synthesize unsaturated fatty acids. Such an un-saturated fatty acid auxotroph could not grow in the ab-sence of an exogenously supplied unsaturated fatty acid, which is presumably required to maintain the membrane in a fluid state at normal growth temperatures. The mutant could grow when either oleic acid (a *cis*-unsaturated, 18-carbon fatty acid) or elaidic acid (the corresponding *trans*-unsaturated, 18-carbon fatty acid) was supplied in the medium, and the fatty acid supplied was incorporated in-tact into membrane phospholipids.

Elaidic acid-enriched membranes were found to undergo a phase change at about 30°C. Below this temperature, the membrane was presumably in a less-fluid, ordered state. By contrast, a membrane containing oleic acid, which intro-duces a greater discontinuity into the membrane structure due to the *cis* configuration of the alkene double bond, underwent a phase transition at a lower temperature, at about 13°C. These facts were used to investigate the de-pendence of lactose permease assembly on membrane fluidity. Bacteria were grown at 37°C with either of the two unsaturated fatty acids as supplement. This temperature is above the membrane phase transition temperatures, and, therefore, the membranes were in a fluid state, regardless of the unsaturated fatty acid incorporated. Subsequently, the cell suspensions were brought to a lower temperature, and isopropylthio-β-galactoside, a potent inducer of the synthesis of β-galactosidase and the lactose transport pro-tein, was added. Synthesis of these two proteins is coordi-nately induced because both are translated from the same messenger RNA (mRNA). Although β-galactosidase is a soluble protein, the lactose permease is an integral mem-

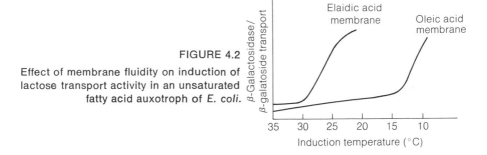

FIGURE 4.2

Effect of membrane fluidity on induction of lactose transport activity in an unsaturated fatty acid auxotroph of *E. coli.*

brane constituent. The ratio of these two enzyme activities measures the relative facility with which the lactose transport protein can be inserted into the membrane in an active form. The results of this experiment are shown in Figure 4.2. If induction of the synthesis of these proteins occurred at a temperature within the normal growth range, which was above the phase transition temperature of a particular membrane, transport activity was maximal and was coordinately induced with β-galactosidase. Below the temperature of the phase transition, transport activity was depressed relative to β-galactosidase activity. This experiment and others led to the conclusion that normal insertion and maximal stability of the lactose permease protein depend on phospholipid fluidity.

Intercellular Junction Formation

A variety of studies has shown that many types of animal cells in close contact form intercellular junctions (*gap junctions*) through which low molecular weight substances can pass. Under optimal conditions, adjacent cells can represent an electric and a metabolic continuum; the introduction of a pulse of current into one cell results in a change in the membrane potential of the adjacent cell. Metabolites, such as nucleotides and sugar phosphates, can be passed from one cell to the other without release of the compounds into the extracellular fluid. These junctions provide a mechanism for direct cross feeding of one cell, deficient for a specific metabolic enzyme, by an adjacent cell that is metabolically competent. The junctions do not, however, allow passage of enzymes or other macromolecules.

Little is known about the mechanism of junctional formation. The maintenance of low intracellular Ca^{2+} concentrations is essential, however; for this reason the Ca^{2+} translocating ATPase, which pumps calcium out of the cell against a concentration gradient, must be active for junction formation to occur. Electron microscopic evidence suggests that junction formation initially involves the rearrangement of integral membrane proteins, which come into close proximity at the site of the future junction. Once formed, junctions are fairly stable and resist physical manipulations; they can be isolated after dissolution of the membrane with appropriate detergent solutions. Although junction formation is undoubtedly dependent on membrane fluidity, the nature of the membrane proteins that give rise to these structures and the forces that cause membrane protein rearrangements have not been extensively characterized.

Myogenesis   Multinucleate skeletal muscle fibers (*myotubes*) arise from the fusion of mononucleate embryonic *myoblasts*. The sequence of events during the myogenic process has been elucidated by extensive ultrastructural, electrophysiologic, and biochemical studies of cultured cells.

Nine-day-old embryonic myoblasts divide in culture every 10 h. Gradually, as the cells divide and age, their cell cycle lengthens, due largely to an increase in the length of one specific phase in the cell cycle, the G_1 phase, which extends from the end of mitosis to the beginning of the phase in which DNA synthesis occurs. Four or five hours before myoblast fusion is initiated, the intracellular concentration of adenosine 3':5'-cyclic monophosphate (cyclic AMP) has been reported to increase many-fold. This increase is followed by an equally rapid loss of cyclic AMP from the cytoplasm, so that the net result is a sudden "spike" of cellular cyclic AMP. Although the exact function of this cyclic AMP spike is not known, it may represent a "trigger," which initiates the fusion process.

For fusion to occur, several conditions must be met. First, the cells must be in an appropriate stage of differentiation. Precisely what this differentiated state entails is not clear, but an appropriate membrane composition or state of phosphorylation is presumably required. Second, the cells must be in the G_1 phase of the cell cycle before the cells can coalesce. Third, both collagen and serum factors (of unknown function) are required; and, finally, the fusion process shows an absolute requirement for Ca^{2+}. If Ca^{2+} is removed from the medium of the myoblast preparation, fusion is postponed; when the ion is reinstated, synchronous cell fusion is observed.

Immature myotubes never undergo cell division, but further fusion occurs with an increase in myotube size. As the myotubes mature, the cellular levels of DNA polymerase rapidly fall, while the activities of muscle-specific proteins are induced to high levels. These proteins include acetylcholinesterase, the acetylcholine receptor protein, creatine phosphokinase, actin, and myosin. Induction of these proteins is not coordinate; for example, creatine phosphokinase can only be induced after fusion has occurred, but the acetylcholine receptor protein can be induced before fusion. The ionic composition of the cytoplasm also changes, probably reflecting the induced synthesis of solute transport components and other membrane proteins. For example, the concentrations of inorganic PO_4 and Mg^{2+} drop, while those of ATP and K^+ increase. The final stage in muscle differentiation appears to be the assembly of functional contractile fibers.

The fusion process has been followed in detail both by electrophysiologic techniques and by light and electron microscopy. Visually, two cells or immature myotubes come into contact and fuse. Electrophysiologic measurements have shown that the young myotubes are electrically coupled shortly after contact is made. A coupling ratio* of 0.2 between adjacent cells is observed, presumably due to intercellular junction formation. At the time of fusion, the apparent value of the membrane potential becomes more negative, and the coupling ratio jumps to 1.0. This process is illustrated in Figure 4.3. The figure also shows an electron micrograph of two muscle cells immediately after fusion. The membrane components in each bilayer seem to randomize at specific sites of contact, possibly at sites where gap junctions had previously formed. Although the fluid nature of the protein components of myotube membranes has recently been established, and membrane fluidity is likely to be a prerequisite for fusion, the molecular details of the process have yet to be elucidated.

Phagocytosis

The process of endocytosis (*phagocytosis*, *pinocytosis*) has been followed and filmed by light microscopy. Phagocytosis of microorganisms and macromolecules by protozoans and phagocytic animal cells involves the introduction of these particles into intracellular vesicles that originate by the folding of the plasma membrane around the material being engulfed. When this process is stimulated, much of the cell surface membrane may appear as intracellular vesicles.

The formation of a food vacuole or a primary phagosome is the first step in intracellular macromolecular digestion. Subsequently, the primary phagosome fuses with a primary lysosome to form a secondary phagosome in which the macromolecular species are exposed to digestive enzymes. The inner lining of the phagosome (previously the outer surface of the plasma membrane) undergoes changes, possibly as a result of the action of these same digestive

---

* The coupling ratio is defined as the ratio of the electric response of the contiguous cell to that of the cell receiving the stimulus.

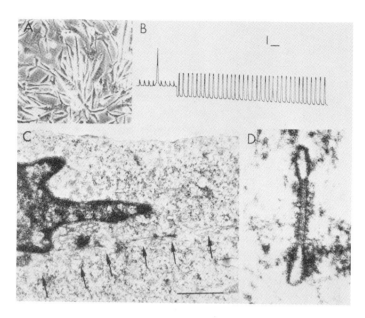

FIGURE 4.3. Illustration of muscle cell fusion. (A) Muscle
cells in the process of fusion as revealed by light microscopy.
(B) An electrophysiologic tracing of the muscle fusion event.
One cell was given periodic pulses of electric current, and the
response was recorded in an adjacent cell. Before fusion, the
coupling ratio was about 0.2. After fusion the coupling ratio
was 1.0. (C) An electron micrograph of two muscle cells
immediately after fusion. (D) View of a vesicle which may have
resulted from the fusion of two segments of the plasma
membranes of adjacent cells.

Reproduced from *Fed. Proc.* 32:1636–1642 (1973), with permission.

enzymes. Consequently, the membrane is altered in physical appearance and becomes increasingly permeable to the
small molecules that are released as the intravesicular
macromolecules are digested. These molecules, then, cross
the vacuole membrane and appear in the cell cytoplasm
where they may be further metabolized. Finally, the metabolites may be utilized for biosynthetic purposes, or they
may enter the mitochondria where they are oxidized for
ATP synthesis.

When the macromolecules in the secondary phagosome
are completely digested, microvesicles may pinch off from
the phagosome. The non-digestable components in the
vacuole are finally released into the extracellular fluid, possibly as a result of a terminal fusion of the vesicle with the
plasma membrane.

Little is known concerning the molecular details of the
phagocytic process. Electron microscopic and biochemical

studies have revealed an organized system of 40–50 Å
diameter *microfilaments*, which underlies and is in association with the plasma membranes of numerous eukaryotic
cells. The microfilaments consist largely of a protein that
resembles actin from striated muscle and, therefore, may
function in a contractile event. Since morphologically similar structures have been found subadjacent to the substrate
attached surfaces of phagocytic cells and in association with
phagocytic vacuoles, it has been suggested that these fila-
ments play a role in the formation of primary phagosomes.
This suggestion has been strengthened by the observation
that such drugs as *cytochalasin B*, which disrupt oriented
bundles of membrane-associated microfilaments, simultaneously inhibit phagocytosis. Unfortunately, a clear picture as to how microfilaments participate in particle-induced
membrane invagination has not yet been formulated.

Secretion   In Chapters 1 and 2 it was noted that proteins are compartmentalized in the Gram-negative bacterial cell. Some soluble proteins are restricted to the cytoplasmic fraction,
whereas others are found only in the periplasm or the extracellular fluid. Given the nature of the secretory process, the
synthesis of alkaline phosphatase, a periplasmic enzyme in
E. coli, has been examined. This protein possesses two subunits, and the dimer requires Zn^{2+} for activity. Genetic and
biochemical studies have led to the conclusion that the
polypeptide is preferentially synthesized on membrane-
bound ribosomes. Since the monomeric subunits apparently appear in the periplasmic space before dimerization
or activation by Zn^{2+}, it has been suggested that peptide
elongation occurs in a process catalyzed by ribosomes
attached to the cell membrane. A vectorial process is envisaged, with release of the completed polypeptides in the
periplasm.

In a like fashion, the first step in protein secretion in a
eukaryotic cell appears to involve synthesis of a polypeptide chain on a membrane surface, that of the rough endoplasmic reticulum. As proposed for secretion of bacterial
periplasmic proteins, the process appears to be vectorial;
the *N*-terminus of the polypeptide chain appears initially
in the lumen of the endoplasmic reticulum rather than in
the cytoplasm (Figure 4.4). Integral surface plasma membrane proteins may be similarly synthesized in the rough
endoplasmic reticulum, the only difference being that the
hydrophobic moiety remains in association with the inner
monolayer of the endoplasmic reticular membrane.

A variety of experiments supports the conclusion that
secretion occurs as shown in Figure 4.4. For example, in
one experiment, puromycin was added to a preparation of

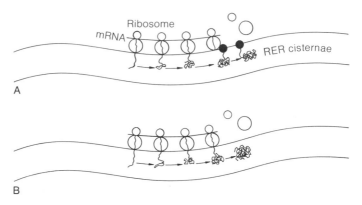

FIGURE 4.4. Proposed mechanism for the synthesis of an
integral membrane protein (A) or a secretory protein (B) by a
membrane-bound ribosome in the rough endoplasmic reticulum
(RER). The figures depict the vectorial translocation of the
polypeptide chain across the endoplasmic reticular membrane
into the lumen during its synthesis.

rough endoplasmic reticular membranes, which were syn-
thesizing a secretory protein *in vitro.* The incomplete polypeptide chains were released from the ribosomes into the
lumen of the reticular vesicles. Moreover, newly synthesized proteins were resistant to digestion by proteolytic
enzymes unless detergents were added to disrupt the
reticular membrane. These results point to a mechanism in
which the protein is released from the ribosome directly
into the cisternae of the vesicles. A process involving transport of the secretory proteins across the reticular membrane
after their release into the cytoplasm appears to be ruled
out.

The second step in eukaryotic secretion involves the
finishing and packaging of the product for export. This is
accomplished in the smooth endoplasmic reticulum and
the Golgi apparatus. Radioactive tracer studies have shown
that a secretory protein is transported from the rough endoplasmic reticulum to the smooth endoplasmic reticulum
and then becomes associated with the Golgi apparatus. If
the protein is to be secreted as a glycoprotein, glycosylation
can occur at any of these sites. In some cases, the initial
glycosylation reaction may occur before elongation of the
polypeptide chain is complete. Extension of the carbohydrate
chains, however, appears to occur principally as the protein passes through the smooth endoplasmic reticulum and
the Golgi apparatus. In the latter organelle, the material is
packaged in small Golgi vesicles, which travel to the periphery of the cell to fuse with the plasma membrane. Fusion

is accompanied by release of the sequestered macromolecules into the extracellular fluid. This sequence of events is represented schematically in Figure 4.5.

The final step in eukaryotic secretion, fusion of vesicle with the plasma membrane, appears to depend on specific interactions between integral proteins in the vesicle and plasma membranes. Although it is likely that more than a single mechanism can account for the release of secretory materials, our knowledge of these processes is sadly incomplete. Only in a few instances has the fusion event been characterized either morphologically or biochemically. Probably the two most extensively characterized exocytotic processes are mucus secretion in the ciliated protozoan, *Tetrahymena*, and neurotransmitter release at mammalian neuromuscular junctions. These two processes are described below.

Mucus is carried to the surface of the *Tetrahymena* cell in mucocysts, large membrane-bounded vesicles, which probably develop from the endoplasmic reticulum. When mature, the mucocysts move to a "rest position" under the plasma membrane. They approach the plasma membrane and eventually fuse with it. Subsequently, the "crystalline" intravesicular material becomes amorphous; the mucocyst, once elongated, becomes spherical, and discharge of mucus occurs. Electron microscopy has been employed after freeze-fracturing of the membrane to reveal morphologic details of the fusion event. The freeze-fracturing technique involves rapidly freezing the membrane preparation of interest at a low temperature and then striking the frozen material with a sharp knife to fracture it. The fracture plane usually passes through the center of a lipid bilayer, disclosing the internal structure of the membrane. Freeze-fracturing has revealed special regions of the plasma and organellar membranes that are structurally differentiated for fusion. The fusion site on the plasma membrane appears as a rosette, a circle of from eight to twelve particles, which

FIGURE 4.5

Intracellular transport of a secretory protein in a eukaryotic cell to the plasma membrane where release of the intravesicular contents occurs. (A) rough endoplasmic reticulum; (B) smooth endoplasmic reticulum; (C) Golgi apparatus; (D) Golgi vesicles; (E) extracellular space.

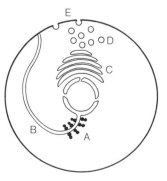

presumably represent deeply embedded integral protein complexes in the membrane. The diameter of each of these intramembraneous particles is about 15 nm, whereas the diameter of the rosette is roughly 65 nm. Rosette particles are visibly larger than most of the other intramembraneous particles (Figure 4.6A). To match this site, an annulus of somewhat smaller protein particles (diameters of about 11 nm) forms at the apical portion of the mucocyst membrane where fusion will eventually occur. (Figure 4.6B). The inner dimensions of this annulus are such that the rosette particles in the plasma membrane appear to fit snugly into the annulus.

The first step in the membrane reorganization process that leads to fusion is the formation of a depression in the center of the rosette—the fusion pocket. In the second step, the pocket deepens and enlarges from a diameter of about 65 nm to a diameter approaching 200 nm, while the

FIGURE 4.6. Intramembrane structures which appear to be involved in the secretion of mucus in *Tetrahymena*. (A) A rosette of intramembraneous particles in the plasma membrane (*arrow*). (B) An annulus of intramembraneous particles in the mucocyst membrane (*arrow*). These structures were revealed by the freeze-fracturing and etching technique.

Reproduced from Satir *et al.* with permission. J. Cell Biology, 56:153 (1973). Copyright © 1973 by the Rockefeller University Press.

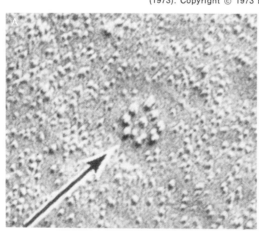

rosette particles spread to the lip and are separated. Finally, as the pocket deepens, the annulus particles from the mucocyst become visible at the lip, indicating completed fusion of the mucocyst and plasma membranes. A schematic representation of this process is depicted in Figure 4.7. This interpretation assumes a stepwise fusion process, in which fusion of the inner face of the plasma membrane with the outer face of the mucocyst membrane disrupts the bilayer structure. Subsequently, dissolution of the two re-

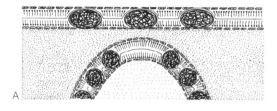

FIGURE 4.7

Schematic representation of the process in which mucocyst and plasma membranes fuse with the release of secretory materials. Figures (A) through (D) show sequential steps in the fusion process.

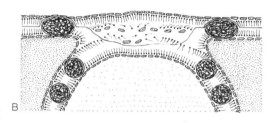

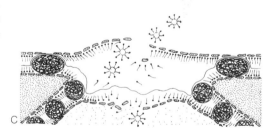

Reproduced from Satir *et al.* with permission.
J. Cell Biology, 56:153 (1973). Copyright ©
1973 by the Rockefeller University Press.

maining monolayers is rapidly followed by reorganization of the membrane constituents so that a cohesive bilayer joins plasma and mucocyst membranes. Migration of both lipid and protein components of the mucocyst membrane into the plasma membrane may serve to enhance the surface area and provide a mechanism for plasma membrane biogenesis.

Eukaryotic cells secrete small molecules as well as large molecules (macromolecules). Among these small molecules are neurotransmitters, such as acetylcholine, which are synthesized and secreted by nerve cells at neuromuscular junctions. The release of neurotransmitters into the synaptic cleft between the nerve and muscle cell membranes provides a mechanism for the chemical transmission of an electric impulse from one excitable cell (the nerve) to a different adjacent one (the muscle cell). Prior to their release into the extracellular fluid, neurotransmitters are sequestered in presynaptic vesicles. The vesicles are synthesized in the Golgi apparatus and then travel down the long axon to the nerve termini. The mechanism by which neurotransmitters (which are synthesized in the nerve termini) become sequestered within the vesicles is not clear, but one of two mechanisms, active transport or group translocation (see Chapter 5), appears likely.

The release of acetylcholine from presynaptic vesicles is greatly stimulated by depolarization of the membrane (the depolarization process will be discussed in Chapter 7). Stimulation of neurotransmitter release may result because depolarization allows Ca^{2+} to flow into the cell from the extracellular fluid, enhancing the cytoplasmic concentration of this ion. Calcium may then, in turn, cause the contraction of actomyosin-like protein complexes, which are thought to be associated with morphologically distinct structures in the plasma and vesicle membranes. This contractile event may trigger neurotransmitter release.

The speculative mechanism proposed above is consistent with much available evidence. For example, an actomyosin-like protein complex has been isolated from nerve endings. This complex has been dissociated into two proteins, one that resembles actin from striated muscle; the other is similar in its physicochemical and enzymatic properties to muscle myosin. Purification of plasma membrane and acetylcholine vesicles from nerve termini has revealed that the actin-like protein is associated with the plasma membrane, whereas the myosin-like protein is primarily associated with the vesicles. Moreover, drugs that inhibit the contractile function of these proteins prevent neurotransmitter release. In view of these observations, it seems reasonable that neurotransmitter release might be initiated when actomyosin complexes associated with the plasma and vesicle membranes contract. Contraction may change the membrane conformation, to produce an opening and allow release of transmitter into the extracellular fluid. It is noteworthy that this mechanism does not require permanent membrane fusion. Possibly, the vesicles dissociate from the plasma membrane and are reutilized for the storage and release of neurotransmitters. This suggestion is in agreement with the observations that neuronal stimulation that promotes neurotransmitter release does not appreciably decrease the number of presynaptic vesicles or enhance the surface area of the plasma membrane at the nerve termini. Whether actomyosin-like proteins are generally involved in endo_ and exocytotic processes has yet to be determined. Conceivably, several distinct mechanisms have evolved for the bulk transport of molecules across biologic membranes.

Selected References

Axline, S. G. and E. P. Reaven. Inhibition of phagocytosis and plasma membrane mobility of the cultivated macrophage by cytochalasin B. Role of subplasmalemmal microfilaments. *J. Cell Biology, 62*:647 (1974).

Berl, S., S. Puszkin, and W. J. Nicklas. Actomyosin-like protein in brain. *Science, 179*:441 (1973).

Braun, V. and K. Hantka. "Biochemistry of bacterial cell envelopes," in *Annual Review of Biochemistry, Vol. 43*. Annual Reviews, Inc., Palo Alto, Ca., 1974, p. 89.

Edidin, M. and D. Fambrough. Fluidity of the surface of cultured muscle fibers. Rapid lateral diffusion of marked surface antigens. *J. Cell Biology, 57*:27 (1973).

Fischback, G. D., D. Fambrough, and P. G. Nelson. Neuron and muscle cell culture. *Federation Proc., 32*:1036 (1973).

Fox, C. F. "Membrane assembly," in *Membrane Molecular Biology* (C. F. Fox and A. D. Keith, eds.). Sinauer Associates, Inc., Stamford, Conn., 1972, p. 345.

Getz, G. S. "Organelle biogenesis," in *Membrane Molecular Biology* (C. F. Fox and A. D. Keith, eds.). Sinauer Associates, Inc., Stamford, Conn., 1972, p. 386.

Leive, L. (ed.). *Membranes and Walls of Bacteria*. Dekker, Inc., New York, 1973.

Loewenstein, W. R. "Transport through membrane junctions," in *The Molecular Basis of Biological Transport* (J. F. Woessner, Jr., and F. Huijing, eds.). Academic Press, New York, 1972.

Novikoff, A. B. and E. Holtzman. *Cells and Organelles*. Holt, Rinehart and Winston, Inc., New York, 1970.

Pitts, J. D. "Direct interactions between animal cells," in *Cell Interactions* (C. J. Silvestri, ed.). Nort, Holland, 1972, p. 277.

Satir, P. and N. B. Gilula. The fine structure of membranes and intercellular communication in insects. *Annual Review of Entomology, Vol. 18*. Annual Reviews, Inc., Palo Alto, Ca., 1973, p. 143.

Satir, B., C. Schooley, and P. Satir. Membrane fusion in a model system. Mucocyst secretion in *Tetrahymena. J. Cell Biology, 56:* 153 (1973).

Schlesinger, M. J., J. A. Reynolds, and S. Schlesinger. Formation and localization of the alkaline phosphatase of *Escherichia coli.* *Ann. N.Y. Acad. Sci., 166*:368 (1969).

Siekevitz, P., G. E. Palade, G. Dallner, I. Ohad, and T. Omura. "The biogenesis of intracellular membranes," in *Organizational Biosynthesis* (H. J. Vogel, J. O. Lampen, and V. Bryson, eds.). Academic Press, Inc., New York, 1967, p. 331.

5 Transmembrane Solute Transport Mechanisms

Interpretation encompasses an infinite number of approaches, none of which is necessarily preferable. While two routes will lead through different regions, the final destination may be the same.

Margaret Rowell

Solute molecules may cross biological membranes as a result of a bulk transport mechanism, as described in Chapter 4, or individual molecules may pass independently through the membrane by one of several different processes. A solute molecule that possesses hydrophobic character and is not excessively large can permeate a lipid bilayer by *simple diffusion*. The process follows diffusion (nonsaturation) kinetics and lacks stereospecificity. The prime requirement for passive permeation is solubility in the hydrophobic phase of the membrane. Long-chain fatty acids, sterols, and a variety of lipophilic drugs fall within this category, and these molecules rapidly enter the cytoplasm.

Lipid bilayers are normally quite impermeable to such hydrophilic solutes as inorganic cations, anions, sugars, and amino acids, all of which are essential for normal cell metabolism. Consequently, catalytic systems that translocate these molecules across the plasma membranes of virtually all living organisms have evolved. Although the mechanism by which a specific solute is transported will differ from organism to organism (Table 5.1), the same fundamental processes are operative throughout the living world. These processes are defined below, and specific examples are considered in detail.

Facilitated diffusion or *mediated transport* is the energy-independent equilibration of solute across a membrane.

Table 5.1   Solute Transport Mechanisms (Tentative Summary)

Solute	Polarity of pump	Proposed mechanism of energy coupling	
		Mammalian intestinal cell	E. coli cell
Na^+	out	1° active transport (Na^+, K^+ ATPase)	2° active transport (?) (coupling)
Ca^{2+}	out[a]	1° active transport (Ca^{2+} ATPase)[b]	2° active transport (coupling)
K^+	in	1° active transport (Na^+, K^+ ATPase)	2° active transport (?) (coupling)
Glucose	in	2° active transport (Na^+ cotransport)	Group translocation (P.T.S.)
Several disaccharides	in	Group translocation (hydrolases)	2° active transport (H^+ cotransport)
Several amino acids	in	2° active transport (Na^+ cotransport)	2° active transport (H^+ cotransport)
Cylic AMP	out	1° active transport (ATP dependent)	2° active transport (coupling)

[a] Ca^{2+} is pumped into mitochondria and certain bacteria.

[b] Or 2° active transport (H^+ or Na^+ counter transport).

Because a membrane protein recognizes the solute molecule and catalyzes its translocation, the process may be *stereospecific* (only one of the two possible optical isomers will be transported), and saturation kinetics* are usually observed. Energy-independent transport may conceivably be of two types: carrier-mediated or channel-mediated. Either process may be responsive to a transmembrane electric potential. A consideration of this last possibility and a discussion of the "gated response" will be included in Chapter 7.

Group translocation is a process in which a substrate is modified by an enzyme that catalyzes the transmembrane translocation step. Energy for the accumulation of the solute is provided by the chemical reaction in which the substrate is altered. The catalytic protein is assumed to be oriented across the membrane in such a way that a substrate is

* The rate of transport by a system that exhibits saturation kinetics increases with increasing substrate concentration until all the carriers are occupied. At this point, transport rate becomes independent of substrate concentration. The substrate concentration at which the half maximal rate of uptake is observed is defined as the Michaelis–Menten constant, K_m. Most transport systems exhibit affinities for their substrates in the range of 10^{-8}–10^{-2}M. A single substrate may be transported by several transport systems with differing affinities.

bound only at one surface, while the product is released at the other.

Active transport involves the coupling of some form of metabolic energy to the transmembrane translocation step so that the unaltered solute is either accumulated within the cytoplasm or extruded from it against a concentration gradient. There appear to be two general types of active transport mechanisms: The first type is directly coupled to chemical energy (ATP) or electric energy (electron flow) and is defined as *primary active transport*. The second type depends on chemiosmotic energy (the membrane potential and/or ion gradients). If the solute molecule is uncharged, its transport must be coupled to ion flow. We shall define this process as *secondary active transport*.

The most extensively characterized mammalian transport systems include (a) intestinal sucrase (group translocation), (b) the Na^+,K^+-translocating ATPase (primary active transport), and (c) the glucose transport system of the brush border membrane of the intestinal mucosa (secondary active transport). Well-characterized bacterial systems include (a) the phosphoenolpyruvate:sugar phosphotransferase system (group translocation), (b) the proton-translocating ATPase (primary active transport), and (c) the lactose permease system in *E. coli* (secondary active transport). Table 5.1 summarizes some properties and the modes of energy coupling of transport systems found in prokaryotic and eukaryotic organisms.

Ion-Transporting Antibiotics

A detailed understanding of biological transport phenomena will ultimately require biochemical dissection and reconstitution of the purified systems in model membranes. Although this goal has been achieved only in a few instances, insight into transport mechanisms has come from studies with ion-transporting antibiotics in artificial lipid bilayers. One such antibiotic, valinomycin, is a cyclic depsipeptide containing a sequence of D-valine, D-hydroxyisovalerate, L-valine, and L-lactate repeated three times. Space-filling molecular models show that the hydrophobic groupings of the hydroxy- and amino acid residues extend outward, conferring on the molecule its exceptional lipid solubility, whereas hydrophilic groups extend inward toward the center of the molecule where they function as ion-complexing agents. Valinomycin transports K^+ with extreme ion selectivity. For example, its preference for K^+ over Na^+ is about 10,000 to 1. By contrast, the gramicidins are linear polypeptides, which are presumed to cyclize in a hydrophobic environment so as to accommodate ions. These antibiotics show little discrimination between Na^+ and K^+, possibly because of their multiple cyclizing options, which allow accommodation of ions of various radii.

Insertion of either antibiotic into an artificial membrane enhances its ionic conductance several orders of magnitude. But, if the membrane is cooled below the temperature at which the solid to liquid phase transition occurs, so that the bilayer loses its fluid character, valinomycin-mediated ion conductance is abolished. By contrast, the rate of gramicidin-mediated transport is relatively insensitive to the state of the lipid phase. The explanation for this interesting observation appears to lie in a fundamental mechanistic difference by which these two ionophores catalyze transmembrane transport. Valinomycin appears to function as a mobile carrier, physically crossing from one side of the bilayer to the other with K^+ complexed to the interior polar residues. Since this movement can only occur in a fluid environment, a membrane transition from liquid to solid phase inhibits this flow of K^+. On the other hand, gramicidin molecules apparently aggregate with one another to form ion-conducting channels. A static structure is thereby formed, which should catalyze ion conduction efficiently in a rigid membrane. These predictions agree with the experimental observations noted above.

Just as ionophores catalyze ion permeation by at least two distinct mechanisms, biological transport proteins may show more than a single mode of action. Channel mediation and carrier mediation of transmembrane solute movement provide two reasonable modes by which the proteins discussed in the sections below may couple energy to translocation. Moreover, the analogy between ionophore functions and biological transport protein function can be carried further. Whereas some ionophores (such as valinomycin and gramicidin) catalyze the electrogenic movement of ions across a membrane, others (nigericin and monensin) catalyze the electroneutral exchange of alkali metals for H^+. Similarly, the anion transport carrier in the red blood cell and transport proteins in the inner mitochondrial membrane appear to catalyze the exchange of one molecular species for another. Thus, ionophores provide intriguing model systems for the normal functioning of a variety of mechanistically different, protein-mediated transmembrane transport processes.

Group Translocation— The Bacterial Phosphotransferase System

Early studies on sugar metabolism in *E. coli* suggested that a novel mechanism was responsible for the transport and phosphorylation of a variety of sugars. The inital step in the utilization of such sugars as glucose, fructose, and mannitol appeared to be phosphorylation, but attempts to demonstrate ATP-dependent phosphorylation reactions that were sufficiently rapid to account for the rates of sugar utilization were unsuccessful. This anomaly was resolved in 1964 when a phosphoenolpyruvate (PEP)-dependent sugar phosphotransferase system (PTS) was discovered. This enzyme system catalyzes the following reaction:

$$PEP + sugar \xrightarrow[Mg^{2+}]{PTS} Sugar\text{-}P + pyruvate$$

where the sugar could be glucose, fructose, mannitol, or any one of a number of other sugars. Detailed biochemical analyses revealed that the enzyme system catalyzing this reaction was complex. Four distinct proteins were required for the phosphorylation of any one sugar. Furthermore, many of the proteins were themselves phosphorylated in the process. Thus, sugar phosphorylation was preceded by a sequence of protein phosphorylation reactions. Phosphate transfer occurred as follows:

$$
\begin{array}{l}
\qquad\qquad\qquad\qquad\qquad\qquad\qquad II^{glc} \\
\qquad\qquad\qquad\qquad\qquad III^{glc}\sim P \longrightarrow Glucose\text{-}6\text{-}P \\
Mg^{2+} \qquad\qquad\qquad\nearrow \qquad\qquad II^{fru} \\
PEP \longrightarrow I\sim P \longrightarrow HPr\sim P \longrightarrow III^{fru}\sim P \longrightarrow Fructose\text{-}1\text{-}P \\
\qquad\qquad\qquad\qquad\qquad\searrow \qquad\qquad II^{mtl} \\
\qquad\qquad\qquad\qquad\qquad III^{mtl}\sim P \longrightarrow Mannitol\text{-}1\text{-}P
\end{array}
$$

According to this scheme, enzyme I and a small heat-stable protein, HPr, are *general* phosphate carrier proteins. These proteins are necessary for the phosphorylation of all sugar substrates of the PTS. Enzymes II and III are the sugar-specific components of the system.* A pair of each of these enzymes appears to be required for the phosphorylation of a particular sugar. Although enzyme III is phosphorylated and, therefore, functions as a phosphate carrier protein, enzyme II is not. Enzyme II is the protein that appears to recognize and bind the sugar. A ternary complex between enzyme III, enzyme II, and sugar must form before phosphate can be transferred from the enzyme III to sugar. Interestingly, all enzymes III isolated are oligomeric proteins consisting of two, three, or four apparently identical subunits. This fact may suggest that the enzymes II, which are always integral membrane proteins, are also oligomeric proteins.

Since the membrane-associated enzymes II are involved in sugar binding, these components are frequently thought of as the sugar-specific carriers or channels. Enzyme I, HPr, and the enzymes III are thought of as energy-coupling proteins. But, experimental results do not support this simplistic view. If these three proteins merely functioned in energy coupling, genetic loss of one of these components

* The original nomenclature for these proteins is more complex than that used here, i.e., see Woessner and Huijing (1972).

by mutation would prevent the intracellular accumulation of the sugar phosphate, but would not prevent carrier function (i.e., facilitated diffusion). In fact, loss of any one of the four proteins involved in the phosphorylation of a particular sugar abolishes transport function except in certain abnormal situations. As a result, one can conclude that phosphorylation of a sugar via the PTS is normally obligatory for sugar entry. Facilitated diffusion cannot be catalyzed by the system. Since the phosphorylation and transport steps are tightly coupled, this process meets the criteria of group translocation.

Group Translocation— Intestinal Disaccharidases

The brush border membrane of the small intestine contains several enzymes that catalyze the hydrolysis of disaccharides. One of these integral membrane proteins has been purified to homogeneity and was found to consist of two enzymatic entities: sucrase, which catalyzes sucrose hydrolysis, and isomaltase, which hydrolyzes a different disaccharide, isomaltose. The catalytic activity of this dimeric protein was studied in an artificial membrane system. The hydrophobic protein was dissolved in a "membrane-forming solution," which consisted of 4% lipid in an organic solvent. A droplet of the solution was brushed over an orifice separating two aqueous chambers. After a period of time, the oil droplet thinned down spontaneously with the formation of a lipid bilayer containing the catalytic protein. Subsequently, radioactive sucrose was placed in one of the two aqueous chambers, which was separated from the other chamber by the bilayer, and the appearance of radioactivity in the second chamber was measured. The purified sucrase–isomaltase complex stimulated transport of radioactive sucrose across the membrane more than 10,000-fold. The radioactive material that appeared in the second chamber, however, was not sucrose, but its constituent monosaccharides: glucose and fructose. Since the enzyme did not stimulate glucose or fructose transport, and since other proteins lacked this catalytic activity, it was concluded that the enzyme was catalyzing vectorial translocation in which substrate was bound at one membrane surface and the hydrolytic products were released at the opposite side. As noted above, such a result would require a highly specific orientation of the catalytic protein across the membrane.

Active Transport Mechanisms in Intestinal Epithelial Cells

Almost by definition, primary and secondary active transport processes are energetically coupled. The energy maintained in an electrochemical gradient created by a primary active transport process is utilized for the active accumulation or extrusion of a solute by secondary active transport. These facts can be illustrated by the transport processes

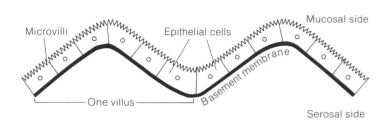

FIGURE 5.1. Schematic diagram of the intestinal mucosa.

that occur in the intestinal epithelial cell lining. The structure of this cell layer is illustrated in Figure 5.1. The epithelial cells that line the intestinal wall have a highly convoluted membrane surface on the mucosal side called the *brush border membrane*. This membrane is rich in carbohydrate (on the external surface) and contains a variety of phosphatases, disaccaridases, and the glucose transport system. The serosal side of this epithelial cell layer is lined with a mat of extracellular fibers (the *basement membrane*), which separates the epithelium from the underlying layers of muscle and connective tissue. Junctions between individual epithelial cells prevent seepage. The membrane on the serosal side of the epithelial cell layer is the site of the Na^+ and K^+ pumps. The Na^+ is pumped out of the cell via the Na^+ pump, whereas the K^+ is pumped into the cell in a coupled process. The electrochemical Na^+ gradient can then be utilized to drive glucose into the cell against a concentration gradient because entry of glucose is tightly coupled to uptake of Na^+ (Figure 5.2).

Early investigations served to characterize the Na^+ and K^+ pumps in a variety of eukaryotic cells. These two pumps appeared to exhibit strict polarity. The Na^+ pump always catalyzed expulsion of Na^+ from the cell, whereas the K^+ pump always catalyzed uptake of K^+. Reversal of this polarity could not be demonstrated. The two pumps apparently required simultaneous operation. In the absence of intracellular Na^+, K^+ uptake was abolished, whereas Na^+ extrusion was only observed in the presence of extracellular K^+. Since both ions were transported against a concentration gradient, the involvement of energy was inferred. The strong diminuation of transport by inhibitors of ATP synthesis led to the conclusion that the energy source might be ATP.

The properties of the monovalent ion pumps led to a search for an ATP hydrolytic enzyme requiring the simultaneous presence of both Na^+ and K^+ for activity. In 1954, such an enzyme was found. Several lines of evidence suggested that this enzyme functioned as the monovalent ion

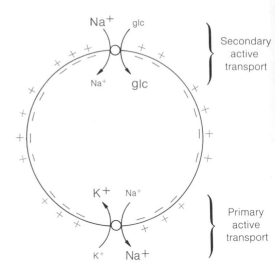

FIGURE 5.2

Schematic diagram of active transport
processes in an intestinal epithelial cell.
The process at the bottom shows the
primary active tansport of Na⁺ and K⁺
by the Na⁺, K⁺-translocating ATPase.
The process at the top illustrates
glucose–Na⁺ cotransport catalyzed by
the glucose carrier.

pump: First, the enzyme hydrolyzed ATP to ADP and inorganic PO_4 in a process that required both Na⁺ and K⁺.
Second, the enzyme was found to be an integral plasma
membrane protein. Cells with high pumping activity showed
high Na⁺, K⁺-dependent ATPase activity and vice versa.
Third, cardiac glycosides, such as ouabain, inhibited ATP
hydrolysis at low concentrations (10^{-6}M). These compounds
were known to be specific inhibitors of the monovalent
cation pump. And, finally, the purified ATPase was functionally reconstituted in artificial phospholipid vesicles.
When ATP was supplied to such a preparation extravesicu-
larly, Na⁺ was pumped into the vesicles, clearly establishing
the ATPase as the Na⁺ transporter.

The enzyme has been purified to homogeneity from
canine renal medulla. The purified complex consists of two
polypeptide chains: a large protein subunit (MW, 135,000)
and a small subunit (MW, 30,000–40,000), the latter being a
sialylglycoprotein. Lipid is tightly bound to the protein
complex. The large protein subunit of the ATPase is now
known to penetrate the membrane. Although the cardiac
glycosides bind to the external surface of this polypeptide
chain, ATP is hydrolyzed at an intracellular site. The function of the small subunit is not known, but it is clearly an
integral component of the outer surface of the bilayer.

The catalytic activity of the ATPase has been extensively
studied. During hydrolytic cleavage of ATP, the β-carboxyl
group of an aspartyl residue in the large subunit is phosphorylated. Whereas protein phosphorylation is stimulated
by Na⁺, hydrolysis of the high-energy protein anhydride
bond depends on K⁺. This observation has led to the suggestion that bound Na⁺ is required for protein phosphoryla-

tion, and that phosphorylation results in a conformational change that causes this ion to transcend the membrane. Similarly, bound K^+ may be required for protein dephosphorylation, which is accompanied by K^+ transport. Since the process is cyclic, interruption of either step would block the entire process. Consequently, Na^+ and K^+ transport are obligatorily coupled.

Antibodies that bind specifically to the inner surface of the large polypeptide chain do not inhibit its catalytic function. Since ATP hydrolysis and ion transport are presumably tightly coupled processes, this result suggests that such large conformational changes as rotation or diffusion of the ion binding sites are not responsible for transport. It has been proposed that the ATPase is normally present as an oligomeric complex with a channel through the center. Possibly minor conformational changes that accompany protein phosphorylation and dephosphorylation are responsible for ion translocation (Figure 5.3).

Most animal cell tissues appear to transport glucose by facilitated diffusion. Active transport of this hexose, however, has been demonstrated in the small intestine, the kidney, and the choroid plexus. Whenever active transport of a sugar or an amino acid has been demonstrated in an animal tissue, a dependence on Na^+ has also been demonstrable. An important question concerns the nature of Na^+ stimulation: Is Na^+ directly involved in energy coupling or does it merely facilitate carrier function?

Several years ago, kinetic analyses of glucose transport in intact intestinal tissue provided evidence for an involvement of Na^+ in energy coupling. Intracellular accumulation of sugar against a concentration gradient was Na^+-dependent and was prevented by energy inhibitors, anaerobic conditions, or the cardiac glycoside inhibitors of the Na^+, K^+-translocating ATPase. In the presence of an energy poison,

FIGURE 5.3. Proposed model for the functioning of the Na^+, K^+ ATPase in ion transport.

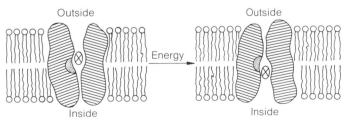

intracellular glucose accumulation could be demonstrated only if the Na$^+$ concentration were lower inside the cells than in the extracellular medium. Moreover, evidence for a Na$^+$ cotransport mechanism was presented: One molecule of Na$^+$ crossed the epithelial cell layer for every molecule of glucose transported.

Studies with isolated chick intestinal epithelial cells have confirmed several of these observations. But, it could also be shown that, in the absence of a metabolic inhibitor, sugar could be pumped into the cell when the Na$^+$ concentration gradient was reversed (i.e., when the Na$^+$ concentration was higher inside the cells than outside). This result clearly showed that the Na$^+$ concentration gradient was not the sole source of energy for sugar accumulation.

Recently, closed vesicles have been isolated from brush border membranes of the rat small intestine. The vesicles were about the size of microvilli (Figure 5.1) and were "right side out." It could be shown that these membrane preparations retained the intact glucose transport system. Here, D-glucose was taken up faster than L-glucose, demonstrating the stereospecificity of the system, and the uptake of D-glucose was stimulated by extravesicular Na$^+$. Since metabolic energy was not available to these vesicles, glucose was not accumulated intravesicularly against a concentration gradient. Transient sugar accumulation, however, was induced with appropriate reagents. A description of these experiments and interpretation of the results are presented below.

Experiment A. When vesicles suspended in a Na$^+$-free buffer were added to an isotonic solution containing 1mM radioactive glucose and 100mM potassium thiocyanate (KSCN), as the only monovalent salt, sugar was taken into the vesicles until the equilibration level was reached. But, if sodium thiocyanate (NaSCN) were substituted for KSCN, entry was more rapid and a twofold accumulation of sugar against a concentration gradient occurred. Subsequently, the level of intravesicular sugar gradually returned to the equilibration level. The only difference between these two experiments was the substitution of Na$^+$ for K$^+$.

The overshoot of glucose uptake appeared to be due to the transient Na$^+$ gradient established when vesicles in Na$^+$-free buffer were added to the NaSCN buffer. Since glucose and Na$^+$ entered the vesicles in a tightly coupled process, the energy in the Na$^+$ gradient provided the driving force for sugar accumulation. If Na$^+$ influx, via the glucose carrier, is an electrogenic process (i.e., the positive charge moves across the membrane without an obligatory neutralization of the resultant charge difference), then a transmembrane electrical potential should be established rapidly (positive inside). This potential will oppose Na^+ influx and thereby inhibit coupled sugar uptake. The lipid-soluble anion, SCN^-, can passively penetrate the membrane and, therefore, should neutralize the electric potential created by Na^+ influx. Destruction of the transmembrane potential should facilitate the flow of Na^+ down its concentration gradient into the vesicles. Thus, glucose overshoot was only observed when NaSCN was present. Replacement of NaSCN with NaCl or Na_2SO_4 abolished the effect.

Experiment B. Vesicles were preloaded with 50 mM K_2SO_4 and then rapidly diluted with medium containing radioactive glucose in which Na_2SO_4 replaced K_2SO_4. The K^+-specific ionophore, valinomycin, was added in one experiment but not the other. In the absence of valinomycin, glucose merely equilibrated across the membrane, but, in its presence, a twofold overshoot was observed. Normally the intestinal membrane exhibits low permeability to K^+. Valinomycin rendered the membranes selectively permeable to this ion and, therefore, allowed K^+ to flow down its concentration gradient out of the vesicles in an electrogenic process. The net flow of positive charges from the vesicle interior created a transmembrane electric potential (negative inside) which provided the driving force for Na^+ influx, and hence for glucose accumulation.

Experiment C. The vesicles were preequilibrated in a buffer containing 50 mM K_2SO_4 at pH 5.25 and were then diluted into a similar solution buffered at pH 7.5 in which 50 mM Na_2SO_4 replaced K_2SO_4. The uptake of radioactive glucose, present in the dilution buffer, was measured in the presence and absence of a potent uncoupler of oxidative phosphorylation, FCCP (carbonyl cyanide *p*-trifluoromethoxy-phenylhydrazone). This compound is thought to act by facilitating proton translocation across the membrane, which is normally only slightly permeable to this ion. In the absence of FCCP, intravesicular radioactive sugar equilibrated with the external medium. FCCP induced a transient overshoot, which was sixfold above the equilibration level. Since FCCP increased the membrane permeability to protons, it should have allowed the electrogenic flow of H^+ across the membrane, down the pH gradient. A membrane potential (negative inside) should have resulted to provide the driving force for Na^+-coupled sugar uptake as described in Experiment B.

In all three experiments, the agent added (SCN^-, valinomycin, or FCCP) allowed coupling of the free energy existing in the Na^+, K^+, or H^+ gradients, respectively, to the transport of glucose against a concentration gradient. That energy coupling occurred via the transmembrane electric field is suggested by the known properties of the agents that promoted sugar accumulation. The results clearly indicate that a positive charge (presumably Na^+) is translocated together with D-glucose by the glucose transport protein across the brush border membrane in an electrophoretic process. This process appears to provide the energy for sugar accumulation.

Energy Interconversion and Active Transport in Bacteria

In the first chapter, structural similarities between mitochondria and bacteria were noted, and an evolutionary relationship was suggested. Careful examination of the membrane components involved in energy transduction in these two biological systems reveals further similarities. For example, both the electron transport chains and the multicomponent coupling factors (the proton-translocating ATPases) appear strikingly similar. With regard to their energy-generating capacity, however, *E. coli* cells are considerably more versatile than mitochondria: they can generate energy either by aerobic substrate oxidation, as do mitochondria, or via anaerobic glycolysis. Figure 5.4 depicts the interconversion of energy as it is thought to occur in *E. coli*. As a result of anaerobic glycolysis, bacteria can synthesize ATP by substrate level phosphorylation. Alternatively, ATP can be synthesized as a result of the reversible reduction and oxidation of electron carriers in the electron transport chain. Formation of ATP in the latter process requires creation of a high energy state, designated "$\sim$" (*squiggle*), and a functional ATPase complex.

The membrane-associated ATPase complex catalyzes a reversible reaction; thus, ATP hydrolysis can drive the formation of "$\sim$," or "$\sim$" can drive ATP synthesis. The nature of "$\sim$" has not been clearly established. It may be a high-energy chemical intermediate or an energized protein con-

FIGURE 5.4. Energy metabolism in *E. coli*. Letters indicate sites of inhibition by: (a) sodium fluoride (NaF); (b) arsenate; (c) dicyclohexylcarbodiimide (DCCD); (d) carbonyl cyanide m-chlorophenylhydrazone (CCCP); (e) potassium cyanide (KCN).

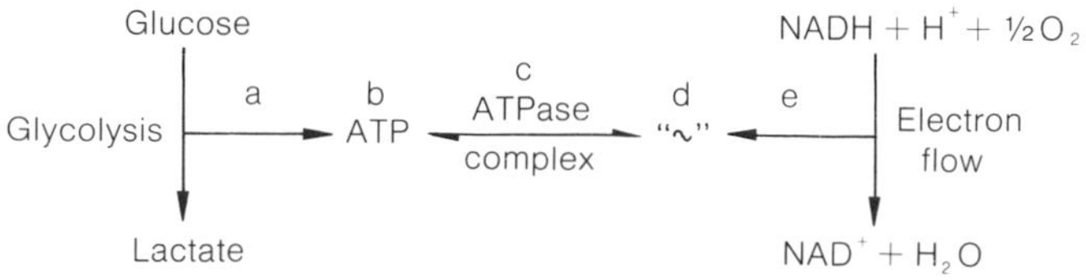

formation. Another theory suggests tha
motic in nature and can be equated to
gradient of H^+. According to this the
"$\sim$," is the sum of the energy storeu ..
brane proton chemical gradient and the membrane ..
potential. Pertinent to this suggestion is the observation
that cytochrome oxidase and the bacterial ATPase apparently catalyze translocation of protons across the membrane
in events that accompany catalytic function. Thus, *cytochrome oxidase* appears to be a *proton-translocating electron carrier*, whereas the *ATPase* is a *proton-translocating
hydrolytic enzyme*. Both membrane-associated protein complexes can be thought of as solute transport carriers. Proton
translocation provides a mechanism for utilizing either
chemical or electrical energy for the generation of chemiosmotic energy. Both ATP hydrolysis and electron flow are
apparently accompanied by proton extrusion in an electrogenic process. Consequently, a pH gradient and a transmembrane electrical potential (negative inside) are generated.
Since these reactions are reversible, the electrochemical
force that results can then be used to synthesize ATP, drive
reverse electron flow, or provide the driving force for the
entry of a positively charged molecule into the cell. It is
this last possibility that we shall be concerned with here.

Lactose is transported across the *E. coli* membrane by an
integral membrane protein, which accumulates the sugar
intracellularly more than 1000-fold against a concentration gradient. The mechanism by which metabolic energy is coupled to transport has been a much debated
question and is still not satisfactorily resolved. But, quantitative measurements have shown that for every molecule
of lactose transported across the membrane, one proton is
transported with the sugar. The lactose transport protein is
a sugar-proton cotransport carrier or channel. As noted
above, a membrane potential (negative inside) could provide the driving force necessary for proton influx and,
hence, for lactose accumulation in a coupled process. Evidence that proton cotransport couples lactose transport to
metabolic energy has been provided by genetic experiments. A mutant bacterial strain, which could still transport lactose rapidly across the membrane by facilitated diffusion (i.e., carrier function was intact), but which could
not appreciably accumulate the sugar against a concentration gradient, was isolated. Thus, the mutant was energy
uncoupled. The genetic defect was shown to be in the gene
that coded for the lactose permease protein. Measurement
of proton fluxes showed that, in the mutant, lactose uptake
was *not* accompanied by uptake of protons. Thus, a genetic

event that resulted in energy uncoupling also uncoupled lactose and H+ cotransport. It was inferred that ion cotransport couples lactose transport to metabolic energy.

Just as the membrane potential appears to provide the driving force for glucose accumulation in intestinal epithelial cells, this source of energy may drive lactose uptake in bacteria. In the absence of electron flow or anaerobic metabolism, a transient transmembrane potential can be produced by suddenly altering the ionic permeability of the bacterial membrane. This was accomplished both in bacterial cells and in membrane vesicles by preloading the cells or vesicles with K+, resuspending them in K+-free buffer, and adding valinomycin to suddenly increase the membrane permeability to this ion. The electrogenic flow of K+ down its concentration gradient created a transient membrane potential. Presumably in response to this potential, lactose was accumulated against a concentration gradient.

The mechanism suggested by these studies is depicted in Figure 5.5. Although there is substantial evidence for this hypothesis, some experimental results suggest that the situation may be more complex. First, an involvement of high-energy PO_4 has been implicated in the transport of some solutes; second, energy appears to influence not only solute transport, but also the binding of some solutes to the carrier; and, third, several observations suggest that many permease proteins lose their catalytic transport activity, even for facilitated diffusion, in the absence of an appropriate energy source. Possibly energy functions in the translocation process in more than a single capacity; even the "shuttling" of a carrier may depend on the energy state of the cell.

FIGURE 5.5

Schematic diagram of active transport processes in the bacterial cell. Protons are pumped out of the cell via either the proton-translocating ATPase or a component of the electron transfer chain, (primary active transport). Lactose can enter the cell in a process coupled to proton entry (secondary active transport).

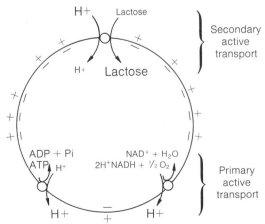

Selected References

Eisenman, G. (ed.). *Membranes, A Series of Advances 2, Lipid Bilayers and Antibiotics.* Marcel Dekker, Inc., New York, 1973.

Goldin, S. M. and S. W. Tong. Reconstitution of active transport catalyzed by the purified Na+, K+-stimulated adenosine triphosphatase from canine renal medulla. *J. Biol. Chem., 249*:5907 (1974).

Harold, F. M. Chemiosmotic interpretation of active transport in bacteria. *Ann. N.Y. Acad. Sci., 227*:297 (1974).

Harold, F. M., K. H. Altendorf, and H. Hirata. Probing membrane transport mechanisms with ionophores. *Ann. N.Y. Acad. Sci., 235*:149 (1974).

Kaback, H. R. Transport studies in bacterial membrane vesicles. *Science, 186*:882 (1974).

Kyte, J. The reaction of Na+, K+ activated adenosine triphosphatase with specific antibodies. *J. Biol. Chem.*, 249:3652 (1974).

Murer, H. and U. Hopfer. "Demonstration of electrogenic Na+-dependent D-glucose transport in intestinal brush border membranes. *Proc. Nat. Acad. Sci., 71*:484 (1974).

Racker, E. (ed.). *Membranes of Mitochondria and Chloroplasts.* Van Nostrand Reinhold Co., New York, 1970.

Semenza, G., A. Quaroni, A. Cogoli, and H. Vogeli. "The small-intestinal sucrase-isomaltase complex: Reconstitution of sucrase-isomaltase-dependent sugar transport with black lipid membranes," in *Protides of the Biological Fluids, 21st Coll., Brugge* (H. Peters, ed.). Pergamon Press, Oxford, 1973.

Singer, S. J. "The molecular organization of membranes," in *Annual Review of Biochemistry, Vol. 43.* Annual Reviews, Inc., Palo Alto, Cal., 1974, p. 805.

Stein, W. D. *The Movement of Molecules across Cell Membranes.* Academic Press, New York, 1967.

Woessner, J. F., Jr., and F. Huijing (eds.). *The Molecular Basis of Biological Transport.* Academic Press, New York, 1972. For excellent reviews on bacterial transport mechanisms.

6 Sensory Perception I: Chemoreception

All you behold, tho' it appear without, it is within—In your imagination, of which this world of mortality is but a shadow.

William Blake

Intriguing questions in biology concern the mechanisms by which chemical and energy stimuli are detected; signals are generated, integrated, and propagated; and responses are elicited. Virtually every living organism responds to any of a variety of stimuli, and, perhaps surprisingly, the membrane-associated molecules, which serve as detectors, have characteristic features. A case in point is photoreception in the mammalian eye, compared with that in the bacterium, *Halobacterium halobium*. Rhodopsin molecules, which capture photons and create (or trigger) an ionic signal, are functional in both organisms. In the mammalian rod cell, ionic conduction triggers a nerve impulse, whereas photoreception in the bacterium functions as an energy generator. In spite of the different end results, the initial processes appear similar.

Chemoreception is another remarkable example. Recent experiments with *E. coli* have established that microorganisms possess chemical receptors in their membranes, which, in some cases, appear to be the proteins involved in transmembrane transport. It has been shown that bacteria possess memory: they respond to chemical gradients *in time* rather than in space. Mammalian responses are similar in these respects. Membrane-associated binding proteins apparently serve as chemical receptors for olfaction and taste, and the response appears to be elicited by chemical gradients in time. Of course, the secondary response is more complex in multicellular organisms, where the integration of complex nerve impulses is required. Nevertheless, the net result is the same: acquisition of or exposure to beneficial substances; movement away from deleterious ones. In this chapter, we shall consider the process of sensory detection, both in primitive and complex organisms. Chapter 7 will be concerned primarily with the cellular transmission mechanisms involved in sensory perception.

Bacterial Chemoreception

Probably the best characterized receptor systems involved in sensory perception are the chemoreceptors in *E. coli.* Chemoreception by a bacterium elicits a response that causes the organism to swim up a gradient of an attractant or down a gradient of a repellent. The target organelle responsible for motility is the bacterial flagellum. It consists of a complex basal body, which is associated with the membrane and cell wall; a long filament, consisting of a single protein organized in a helical array; and a "hook," which links the filament to the basal body. Recent experiments have shown that the filament, and presumably a portion of the basal body to which it is attached, can rotate either clockwise or counterclockwise. As we shall see, the frequency with which this rotational motion changes direction determines the chemotactic response.

The fact that microorganisms can sense and respond to chemical agents was recognized in the seventeenth century when it was noted that swarms of bacteria collect around bits of food. Systematic observational analyses of these accumulations, recorded in Germany between 1880 and 1920, revealed that bacteria exhibit positive tactic behavior toward beneficial gases, nutrients, and energy sources but avoid strong acids and hydrophobic substances. Moreover, a bacterium can respond both positively (at low concentrations) and negatively (at high concentrations) to a single compound, allowing the organism to seek an optimal concentration.

Studies concerned with the effects of nonphysiologic chemicals on bacterial tactic behavior have shown that a positive chemotactic response can be selectively disabled without inhibiting motility. Appropriate concentrations of an anesthetic, such as diethylether or chloroform, caused bacteria to swim past a gradient of a chemical attractant without noticing it. Additionally, it was found that bacteria respond to potent narcotics, such as cocaine, in a rather peculiar way: An organism caught within the field of influence of this drug initiates a rhythmic shuttling motion, swimming back and forth at normal speed, moving about a body length in each direction. Subsequently, the cell slows down, wriggling irregularly, and, eventually, motion ceases altogether. It seems that bacteria, like higher organisms, are

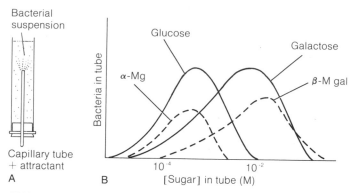

FIGURE 6.1. Capillary tube method for measuring chemotaxis.
α-Methylglucoside (α-Mg); β-Methylgalactoside (β-Mgal).

responsive to anesthetics, an observation that suggests that they might possess a "nervous system" that controls their tactic behavior. In view of the known modes of action of anesthetics (Chapter 3), one would suspect that action potentials might mediate tactic responses in prokaryotic as well as eukaryotic organisms.

Several quantitative methods have been developed for estimating a chemotactic response. One such method is illustrated in Figure 6.1A. A dilute suspension of motile bacteria in a salts-containing medium is placed in a small vessel stoppered at one end. The open end of a capillary tube (1 μl total capacity), filled with a chemical attractant or repellent and sealed at one end, is inserted into the bacterial suspension through the hole in the stopper. In the absence of an attractant, or in the presence of a uniform concentration of the attractant, a certain low number of bacteria migrate into the capillary tube as a result of random motion. The bacteria show no chemotactic response under these conditions. If the concentration of the attractant is higher in the capillary tube than in the medium, some of the attractant in the capillary tube will diffuse out into the medium and set up a concentration gradient. It is the *gradient* of a compound, not the compound itself, the organisms sense. When a positive gradient of an attractant is created, bacteria swim into the capillary tube. After an appropriate period of time, the capillary tube can be removed and the bacteria within it are counted. The number of cells in the tube is a measure of the chemotactic response.

Among known bacterial chemotactic attractants are sugars, amino acids, and organic acids. Repellents include phenolic compounds, hydrophobic molecules, and certain other compounds possibly deleterious to the cell. Responses to gradients of simple sugars have been most carefully investigated. When various sugars were tested for their capacities to elicit a chemotactic response, the results shown in Figure 6.1B were obtained. Glucose and α-methylglucoside both served as effective attractants. These two sugars are substrates of the phosphoenolpyruvate:glucose phosphotransferase system (Chapter 5). Similarly, galactose and β-methylgalactoside, which are substrates of an active transport system, elicited chemotactic responses. By contrast, lactose, a disaccharide of glucose and galactose, which is transported and metabolized by an independent catabolic enzyme system, was without effect. These experiments provide important information about the chemotactic process. Since α-methylglucoside and β-methylgalactoside cannot be metabolized after they are accumulated intracellularly, it can be concluded that a metabolite or an energy source derived from the sugars does not elicit the response. Moreover, lactose, which is hydrolyzed to glucose and galactose intracellularly, was ineffective as an attractant, suggesting that the receptor molecules that detect the attractant must be located on the external surface of the cell.

To characterize the various chemotactic systems that recognize attractants, the inhibitory effects of potential substrates were studied. It was known that bacteria respond to the *ratio* of the different concentrations rather than to the absolute difference. They exhibit a *linear response* to an exponential increase in the attractant concentration. This fact allowed the determination of substrate specificities of individual receptors, and it further allowed investigators to ascertain whether two chemotactic sugars were recognized by the same or different receptors. Figure 6.2 illustrates the technique. An attractant (glucose in Experiment A; galactose in Experiment B) is placed in the capillary tube at a low concentration (0.001M). The inhibitor to be tested (α-methylglucoside in the experiments depicted in Figure 6.2) is placed in the capillary tube *and* in the bacterial suspension at much higher concentration (0.1M). If the chemo-

FIGURE 6.2. Competitive inhibition of chemotaxis. α-Methylglucoside (α-Mg).

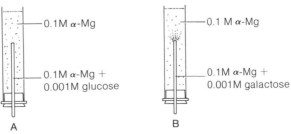

0.1M α-Mg

0.1M α-Mg + 0.001M glucose

0.1 M α-Mg

0.1M α-Mg + 0.001M galactose

A   B

tactic receptor specific for the attractant in the capillary tube (glucose or galactose) does not recognize the inhibitory sugar (α-methylglucoside), the attractant concentration gradient will be from 0 to 0.001M. By contrast, if the receptor *does* recognize the glucoside as a substrate, the effective concentration gradient of attractant will be from 0.100M to 0.101M. Since the concentration ratio is nearly equal to one in the latter case, no chemotactic response would be expected. For the experiments shown in Figure 6.2, a chemotactic response was observed in B, but not A, suggesting that glucose and α-methylglucoside were recognized by a single receptor molecule but galactose was recognized by a different receptor.

Similar experiments have been conducted with numerous attractants. Table 6.1 is a summary of the data that serve to characterize a few receptor systems. Particularly worthy of note is the observation that the specificities and induction properties of the chemotactic systems are the same as those of the corresponding transport systems discussed in Chapter 5. This correlation would suggest that a component of each of the transport systems serves as a chemotactic receptor; that is, the protein is bifunctional. A large body of experimental evidence now supports this conclusion.

Most transport systems in bacteria seem to consist of several distinct proteins. This fact has been unequivocally demonstrated for the sugar-transporting phosphotransferase system in which four proteins must function for the translocation of a particular sugar to occur (Chapter 5). The multicomponent nature of the galactose and maltose transport systems (active transport systems) has also been established. If any one of the phosphotransferase components

Table 6.1 Bacterial Transport and Chemotaxis Systems[a]

System[a]	Substrates	Inducible	Chemotaxis in transport mutants	Transport in chemotaxis mutants
Glucose	Glucose Methyl α-glucoside	−	−	−
Mannitol	Mannitol	+	−	−
Galactose	Galactose Methyl β-galactoside Fucose	+	+ or −	+ or −
Maltose	Maltose	+	+ or −	+ or −

[a] Glucose and mannitol are PTS substrates; galactose and maltose are substrates of active transport systems.

(enzyme I, HPr, enzyme II, or enzyme III) is rendered non-functional, both transport and chemotaxis are blocked. This suggests that passage of sugar through the membrane may be required before a substrate of the phosphotransferase system will elicit a chemotactic response. By contrast, only *one* of the components of each of the active transport systems appears to be required for chemotaxis. The genetic loss of the other components of these transport systems does not appreciably alter the chemotactic response. Similarly, other components of the corresponding chemotactic systems are not required for normal transport function. The one gene product that is essential to both transport and chemotaxis is a sugar-specific binding protein, which confers upon both the chemotactic and transport systems their substrate specificities. Thus, a maltose-specific binding protein is an essential component of both the transport and chemotactic systems for maltose, whereas a galactose binding protein serves as the receptor for both transport and chemotaxis of galactose and β-methylgalactoside. These proteins have been shown to occur in the periplasmic space of the bacterial cell, possibly in association with the cytoplasmic membrane. These presumed receptor molecules are therefore external to the cytoplasmic membrane, as was predicted.

As noted above, chemoreception may be mediated by the phosphotransferase system in a process that requires transmembrane permeation of the attractant, whereas chemoreception of sugar substrates of active transport systems does not appear to require translocation. These observations suggest that more than a single mechanism of chemoreception may exist in bacteria. Just as a multiplicity of mechanisms has evolved for transporting solute molecules across biologic membranes, a multiplicity of chemoreception mechanisms may have evolved in parallel.

Bacteria respond to gradients of chemicals by moving toward the source of an attractant and away from the source of a repellent. But what does the response entail, and how is it elicited? Observations under the microscope have shown that the chemotactic response does not involve a change in the velocity of bacterial swimming; nor are bacteria capable of directly responding to a chemical gradient by veering in the appropriate direction. Instead, chemotaxis serves to coordinate motility. In the absence of a chemical gradient, a bacterium swims in a straight line for a period of time; then, suddenly, linear motion is interrupted while the organism "tumbles" or "twiddles" for a split second before it resumes linear motion in a new and randomly chosen direction. It is the frequency of tumbling that is influenced by a gradient of a chemical attractant or repellent. If the bacterium is moving up an attractant gradient, its tumbling motion is suppressed. If it is moving *down* the same gradient, the frequency of tumbling increases. Interestingly, this behavior appears to correlate with the direction of flagellar rotation. Counterclockwise rotation results in smooth linear motion, whereas clockwise flagellar rotation results in tumbling. It would therefore appear that the chemotactic response, mediated by a specific protein in the bacterial plasma membrane, must ultimately control the direction of flagellar rotation. Although the transmission mechanism relating reception to response is not yet understood, it has been shown that the driving force for bacterial motility is probably provided by the membrane potential, a chemiosmotic form of energy. It is possible, therefore, to envisage an ionic basis for flagellar rotation, and hence, for regulation of the chemotactic response.

Bacterial cells possess memory: they respond to chemogradients *in time* rather than in space. This fact was elegantly demonstrated using the "temporal gradient apparatus" shown in Figure 6.3. Bottles A and B contained medium, but only bottle B contained bacteria (2×10^7 cells/ml). Attractant was added to bottle A or to bottle B or to both, depending on whether a positive, negative, or zero concentration gradient was desired. The two bottles were connected via a peristaltic pump to the inlets of the rapid-mixing device. Residence time in the mixing tube was about 0.2 s, and observation commenced about 1 s after flow was stopped. This apparatus allowed bacterial motility to be studied after the organisms had been subjected to a sudden change from one uniform concentration of attractant to another. A sudden decrease in attractant concentration

FIGURE 6.3. Diagram of temporal gradient apparatus.

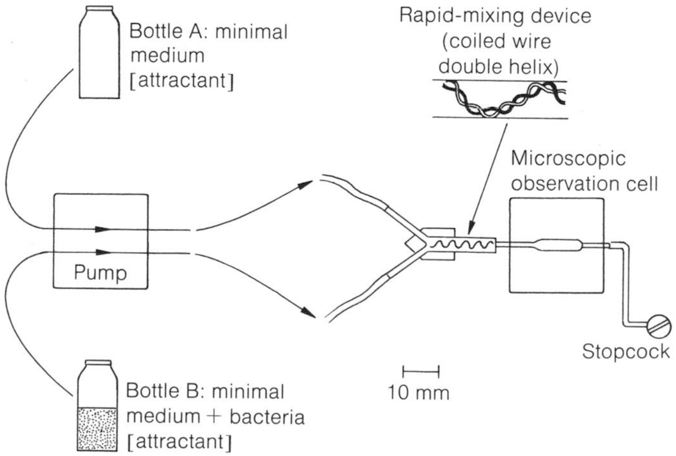

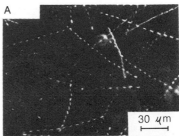

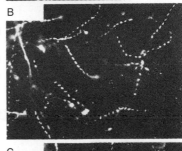

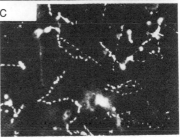

FIGURE 6.4

Motility tracks of S. *typhimurium*, taken in the time interval 2→7 seconds after subjecting the bacteria to a sudden change in attractant concentration in the temporal gradient apparatus. (A) $\Delta C > 0$; smooth, linear trajectories after increasing the attractant concentration. (B) $\Delta C = 0$; normal motility. (C) $\Delta C < 0$; poor coordination after a decrease in attractant concentration.

elicited the tumbling response observed with spatial gradients, whereas an increase elicited supercoordinated swimming. This fact is illustrated in Figure 6.4. Increasing the attractant concentration decreased the frequency of tumbling. In some cases, the response lasted as long as 5 min before normal motion was restored. This experiment demonstrated that chemotaxis is achieved by modulation of the incidence of tumbling, both above and below its steady-state value. The detection of a spatial gradient by bacteria, therefore, apparently involves detection of a temporal gradient experienced as a result of movement through space.

Chemotactic Responses of Solitary Eukaryotic Cells

A variety of eukaryotic cells has been shown to respond tactically to environmental chemicals. Depending on the biologic system, the chemotactic response may function in nutrition, reproduction, differentiation, or defense. A few of the better-characterized examples will be described.

Ciliated protozoa respond to a variety of attractants (amino acids, sugars, and organic acids) as well as to chemical repellents (hydrophobic compounds and acids). Moreover, they may respond either positively or negatively to a single compound, depending on its absolute concentration. *Paramecium* species, for example, swim up a gradient of acetic acid when the concentration is low, but down the gradient when the concentration is high. Similar behavior toward dissolved gases has been noted. These marvelous organisms have apparently evolved complex mechanisms that allow them to select optimal concentrations of these agents. Here, as with bacteria, the chemotactic response appears to serve the cell in acquisition of nutrients and avoidance of deleterious compounds.

The chemotactic behavior of cellular slime molds has been extensively studied. Solitary amoebae of several slime mold species both emit and receive the chemical signal that allows thousands of cells to aggregate and form a multicellular "organism." The orientation of cells toward an aggregation center and the formation of streams of actively moving amoebae are shown in Figure 6.5. This chemotactic

FIGURE 6.5.   Aggregation of the cellular slime mold, *Dictyostelium discoideum*. Cells are migrating inward toward the aggregation center via continuous streams. In these organisms, a center can attract cells only for a distance of about 0.1 mm; yet a stream of cells may radiate 10 mm from a center. This suggests that the cells in the stream must produce cyclic AMP.

Courtesy Dr. P. Farnsworth, University of California at San Diego.

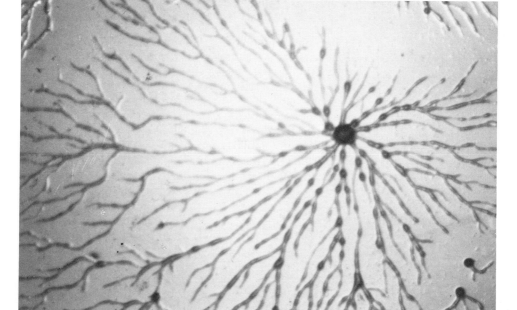

response is the first in a series of events, which eventually leads to differentiation of distinct cell types and the erection of a fruiting body. Production of the chemotactic agent, which has been identified as adenosine $3':5'$-monophosphate (cyclic AMP), is initiated upon exhaustion of a nutrient source by the cell population. In this case, the chemotactic response functions in species preservation by initiating the formation of differentiated spores.

One of the first investigations of eukaryotic chemotactic behavior was described in 1884 by Pfeffer. In these studies, bracken fern spermatozoa were shown to accumulate at the female archegonium. In subsequent studies, the involvement of a chemotactic process was established, and malic acid was identified as the probable attractant. Sperm from a wide variety of plant and animal species are now known to show chemotactic responses that facilitate fertilization.

Certain mobile cells of multicellular organisms respond to chemical stimuli. For example, polymorphonuclear leukocytes, one class of white blood cells, which are phagocytic, exhibit amoeboid motility and sense substances that arise during infection. In response to the substances produced, they can migrate from the bloodstream to a focus of inflammation. Once at a site of infection, these cells will ingest and destroy the infectious agent. *In vitro* studies have shown that a variety of macromolecular agents (bacteria, yeast, carbohydrates, and antigen–antibody complexes) elicit the tactic response. The compounds that function as true chemotactic attractants may be relatively small products released from these agents. But, the chemotactic response of leukocytes is apparently mediated by serum factors, including components of the complement system (one part of the organisms' immunologic defense system). It seems clear that leukocyte chemotaxis functions in organismal defense as well as cellular nutrition.

The Chemical Regulation of Insect Behavior

Chemical messengers, called pheromones, frequently serve to integrate the behavior of individuals or of populations of individuals within a particular animal species. For example, ants use chemical cues to mark trails to food sources, whereas vertebrates and invertebrates alike employ olfactory signals to locate their mates.

One of the most carefully studied pheromone–chemoreceptor systems is that found in the silk moth, *Bombyx mori*. Bombykol, an unsaturated fatty alcohol, is a sex attractant produced by the female and detected by chemoreceptors on the antennae of the male. Of the thousands of antennal receptors available, about one-half respond specifically to bombykol. Although the female's antennae are insensitive to this chemical, those of the male are so sensitive that adsorption of one molecule of the pheromone to the receptor surface will elicit a nerve impulse. When a male detects the volatile bombykol molecule, he proceeds ·upwind unless the scent is lost, in which case he initiates random motion. When the concentration of bombykol reaches a certain critical value, behavioral changes can be observed, and the male begins to track the female by an odor gradient.

The chemoreceptors involved are located in antennal bristles, the spacing and diameter of which create a molecular sieve. Adsorption of a molecule will depend upon the areodynamic conditions, the temperature, the concentration of the pheromone, and the physicochemical properties of the molecule and the bristle. The preferential adsorption of bombykol to chemoreceptor. hairs is indicated by studies with radioactively labeled bombykol in which 80% or more of the label was found on the chemoreceptor hairs, even though the hairs constitute only 13% of the antennal surface area.

Once adsorbed, the pheromone must penetrate the bristle to the sensory nerves enclosed within the shaft of the bristle. The outer layer of the bristle is formed of cuticle, a hardened matrix of chitin, protein, and lipid, which serves as the exoskeleton of the animal. Bombykol molecules do not penetrate the cuticle but probably pass through pores found along the length of the shaft. Electron micrographs of sections of the antennal hairs show that inside the pores are tiny fluid-surrounded tubules, about 30 A in diameter, which lead to the sensory nerves. The bombykol molecules presumably penetrate the pores and pass along the tubules or within the liquor to the sensory nerve surface. By testing the effect of other organic compounds, it has been determined that lipid solubility is an important property such that increasing carbon chain length decreases the threshold concentration for stimulation. Thus, the factors affecting diffusion of the stimulating molecules to the nerve account in part for the specificity of the chemoreceptor.

Because radioactive-labeled bombykol has never been detected by autoradiography inside the nerve, activation of the nerve is presumed to occur at the membrane surface by means of interactions with an acceptor molecule associated with the membrane. From structure–function studies of the stereochemistry of bombykol, it is apparent that the terminal hydroxyl group and the configuration due to the cis and trans double bonds are vital to the biologic activity of the pheromone. Binding to the acceptor is thought to elicit a generator potential that induces an action potential near the cell body of the nerve. Although one molecule of bombykol can elicit a nerve impulse, a couple of hundred molecules are required to effect the behavioral response. The nature of the complex integrative process in the brain, which produces the tracking response, is completely unknown.

The Chemical Senses of Mammals

Chemical cues serve multicellular organisms in several distinct capacities: in nutrition, reproduction, protection, and communication. As a consequence of the frequent use of chemoreception for nutrient detection and the regulation of behavior, virtually all higher organisms have evolved differentiated organs of gustation (taste) and olfaction (smell). In vertebrates, taste buds are the organs of gustation. In man, these organs are confined to the oral cavity, principally to the tongue, but, in other species, they may cover the head and much of the body surface. The taste bud is a bulblike structure about 50 μ in diameter, and it contains about 50 epithelial cells, which are differentiated for chemoreception and signal transmission. Surprisingly, in man, taste buds are not located on the surface of the tongue; instead, a narrow "taste pore" leads from the tissue surface to the bud. Microvilli of bud epithelial cells, which are about 2 μ long and 0.2 μ wide, contain the specific chemoreceptor proteins. The microvilli project into the taste pore and thereby come in contact with the saliva, which contains the nutrients. Individual receptor cells may be differentiated to detect a compound possessing one of the four distinct taste qualities recognized by man: sour, sweet, bitter, and salty. Some investigators believe that a different receptor protein recognizes each of these classes of compounds. In fact, a protein that binds sweet compounds has been isolated, and, on the basis of its binding specificity, it has been implicated in the chemoreception process. This protein may be functionally similar to the bacterial-binding proteins, which initiate the chemotactic response of microorganisms to nutrients.

Olfactory receptor cells are bipolar sensory neurons found in the posterior region of the nasal cavity. Their apical segment is enlarged to form the olfactory vesicle (2 to 3 μ in diameter), from which numerous cilia and microvilli extend. Specific receptor proteins, each with different binding specificities, have been postulated, but neither the numbers nor the specificities of these proteins have been defined.

Stimulation of either gustatory or olfactory receptor cells results in an alteration of the ionic conductance of the membrane such that the membrane potential becomes more positive (depolarization; see Chapter 7). Depolarization, in turn, stimulates synaptic activity, which gives rise to a nerve impulse along the appropriate cranial nerve. Principles applicable to the cellular response and transmission mechanisms in sensory perception will be the subject of the next chapter.

Selected   Adler, J. Chemoreceptors in bacteria. *Science, 166*:1588 (1969).
References Adler, J. and W. Epstein. Phosphotransferase-system enzymes as chemoreceptors for certain sugars in *Escherichia coli* chemotaxis. *Proc. Nat. Acad. Sci., 71*:2895 (1974).

Beidler, L. (ed.). *Handbook of Sensory Physiology, Vol. IV, Chemical Senses.* Springer-Verlag, Berlin, 1971.

Berg, H. C. "Chemotaxis in bacteria," in *Annual Review of Biophysics and Bioengineering, Vol. 4.* Annual Reviews, Inc., Palo Alto, Cal., 1975, pp. 119–136.

Ciba Foundation, *Locomotion of Tissue Cells, Ciba Foundation Symposium 14.* Associated Scientific Pub., Amsterdam, 1973.

Davies, J. T. "Olfactory theories," in *Handbook of Sensory Physiology, Vol. IV* (part 1) (L. Beidler, ed.). Springer-Verlag, New York, 1971, pp. 322–350.

Hodgson, E. S. "Chemoreception," in *The Physiology of Insects, Vol. II*, 2nd Ed. (M. Rockstein, ed.). Academic Press, New York, 1974, pp. 127–164.

Macnab, R. M. and D. E. Koshland. The gradient sensing mechanism in bacterial chemotaxis. *Proc. Nat. Acad. Sci., 69*:2509 (1972).

Schneider, D. (ed.). *Olfaction and Taste IV, Proceedings of The Fourth International Symposium.* Wissenschaftliche Verlagsgesellschaft MBH, Stuttgart, 1972.

Shaffer, B. M. "The acrasina," in *Advances in Morphogenesis, Vol. 2* (M. Abercrombie and J. Brachet, eds.). Academic Press, New York, 1962, pp. 109–182.

Silverman, M. and M. Simon. Flagellar rotation and the mechanism of bacterial motility. *Nature, 249*:73 (1974).

Trinkaus, J. P. *Cells Into Organs, The Forces That Shape The Embryo.* Prentice-Hall, Inc., Englewood Cliffs, N.J., 1969.

7 Sensory Perception II: Transmission Mechanisms

Any appearance whatever present themselves,
not only when its object stimulates a sense,
but also when the sense by itself alone is stimulated,
provided only it be stimulated in the same manner
as it is by the object.

Aristotle

When a bacterium senses a gradient of a chemical attractant through a chemoreceptor, a signal that influences the direction of flagellar rotation must be transmitted to the flagellar basal body. The nature of the signal elicited by the chemoreceptor is not known; nor are the factors controlling the rotational direction of the bacterial flagellar filament understood. But, electrophysiologic studies with much larger eukaryotic cells have provided important insights into the transmission mechanisms that allow sensory receptors to control the activities of the responding organs or organelles. In all well-documented instances, the responses to sensory input are elicited by bioelectric signals. In this chapter, the nature of these signals and the responses elicited are examined.

The Resting Membrane Potential and Action Potentials

An unequal distribution of charged mobile molecules across a semipermeable barrier can give rise to a transmembrane electric potential. The magnitude of this potential will depend on the selective permeabilities of the barrier to the ionic species present. The numerical value of the membrane potential is approximated by the following equation:*

* The abbreviations used are: V_m, the transmembrane electric potential; E_{Na+}, the equilibrium electric potential of Na^+; R, the gas constant; T, the absolute temperature; F, the Faraday; P_{Na}^+, the permeability of the membrane to Na^+; $[Na^+]$, the sodium concentration.

$$V_m = \frac{2.3\ RT}{F} \log_{10} \left[\frac{\Sigma P_c\ [C]\ out + \Sigma P_a\ [A]\ in}{\Sigma P_c\ [C]\ in\ + \Sigma P_a\ [A]\ out} \right]$$

for univalent cations (C) and anions (A).

When a membrane is selectively permeable to one ion (C), this equation reduces to the familiar Nernst equation:

$$V_m = E_c = \frac{2.3\ RT}{F} \log_{10} \frac{[C]\ out}{[C]\ in} = (60\ mV)\ \log_{10} \frac{[C]\ out}{[C]\ in}$$

What these equations relate is the fact that as the permeability of the membrane to a particular ion becomes larger, the membrane potential approaches the equilibrium electric potential of that ion. This relation provides the ionic bases for resting potentials across biological membranes and the action potentials that occur in electrically excitable cells.

In the resting state, the nerve cell membrane is selectively permeable to K^+, whereas permeability to Na^+ is relatively low. Thus, the membrane resting potential will be near the K^+ equilibrium potential and is negative inside. Conceptually, it is easy to understand why this is so. In a nerve cell, Na^+ is actively pumped out of the cell, while K^+ is accumulated intracellularly in a reaction that depends on the Na^+, K^+-translocating ATPase and requires the cleavage of the terminal pyrophosphate bond in ATP. This pumping action creates concentration gradients of Na^+ and K^+, with the Na^+ concentration higher outside the cell than inside it, and the K^+ concentration higher inside than outside. Consequently, Na^+ will tend to diffuse down its concentration gradient into the cell, while K^+ will tend to flow down its concentration gradient out of the cell. Because the nerve membrane is relatively impermeable to Na^+, its passage through the membrane is impeded. But because the membrane is much more permeable to K^+, this ion will flow down its concentration gradient into the extracellular fluid. The electrogenic flow of K^+ will not be able to continue for a long period of time because the unidirectional passage of this charged species will create charge imbalance: there will be an excess of positive charges in the extracellular fluid and a deficiency of positive charges in the cytoplasm. In other words, the cytoplasm will be negative relative to the extracellular fluid. This net charge imbalance will set up a transmembrane electric potential that will oppose the K^+ chemical concentration gradient. At equilibrium, the energy in the electric potential will be equal in magnitude (but opposite in sign) to the energy in the K^+ chemical gradient. The *net* energy in the K^+ electrochemical gradient (the sum of the K^+ chemical and electric gradients) will be equal to zero, and, hence, an equilibrium situation will be attained in which no net flow of charge will occur.

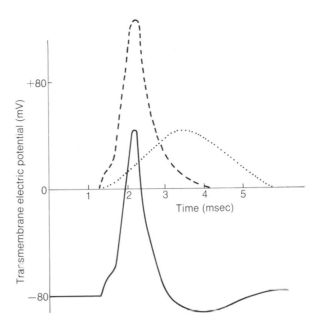

FIGURE 7.1. Typical action potential, which results when a nerve cell is electrically excited. The dashed line indicates Na⁺ conductance, the dotted line shows K⁺ conductance (conductance in arbitrary units).

It was noted above that, in the resting state, the nerve membrane is selectively permeable to K^+ and relatively impermeable to Na^+ and Cl^-. But, the permeabilities (or conductances) of the membrane to both Na^+ and K^+ depend on the membrane potential. Depolarization of the membrane (which renders the membrane potential more positive inside) increases the Na^+ and K^+ permeabilities, whereas hyperpolarization (which renders the membrane potential more negative inside) decreases these permeabilities. Since entering Na^+ depolarizes the cell, and K^+ efflux hyperpolarizes the cell, the *threshold*, above which an action potential will result, is the point at which the velocity of Na^+ influx is equal to the velocity of K^+ efflux.* Above this threshold value, Na^+ entry overpowers other ion movements, until the cytoplasm acquires sufficient positive character to balance the Na^+ chemical potential gradient. This level is never actually attained; first, because the opening of the Na^+

* This statement assumes an involvement of Na^+ and K^+ only and is, therefore, an approximation to the true situation. Other ions (e.g., Ca^{2+}) may be involved to a lesser extent.

channel is only transient, and, second, because it is rapidly followed by an increase in the K^+ permeability. These selective permeabilities to Na^+ and K^+, which temporarily vary independently of one another, result in the familiar action potential shown in Figure 7.1.

Our understanding of the cellular electrical activities that dominate the field of neurophysiologic behavior is considerable at the conceptual level but meager at the molecular level. For example, in spite of exhaustive studies, we have little information about the nature of the potential-sensitive channels responsible for action potentials. The most significant information we do have has come from studies with artificial membranes into which isolated bacterial proteins have been inserted. These studies will be described below.

Excitability-Inducing Material

Artificial lipid bilayers (black lipid membranes) can be constructed across a small orifice separating two chambers filled with an aqueous electrolyte solution (Chapter 5). By inserting electrodes into the two chambers, the electric potential across the membrane can be measured, and the transmembrane flow of ionic current can be quantitated. When the current is measured as a function of voltage, a linear response is observed (Figure 7.2). If valinomycin molecules are inserted into the membrane, the K^+ conductance can increase several orders of magnitude, but a linear response to the transmembrane electric potential is still observed (Figure 7.2). The linear response implies that the conduction properties of this ionophorous antibiotic are independent of the membrane potential.

Several years ago, a protein preparation was obtained from a bacterial source (and later from other biological materials) that conducted ions across artificial lipid bilayer membranes as shown in Figure 7.2. As the transmembrane electric potential was gradually increased from zero, a linear increase in conductance was initially observed. But, as the membrane potential was increased further, the conductance leveled off and then dropped dramatically to a lower value. Subsequently, the differential increase in conductance with potential remained approximately constant, but the slope of the plot was less than it was at a lower potential. Brief reflection on this behavior revealed an analogy between the response to a potential change of the Na^+ channel in a nerve membrane and this protein in a synthetic bilayer. In both cases, depolarization of the membrane from a large (negative) value enhanced the ionic conductance. Because the protein conferred upon a bilayer the properties of an excitable membrane, it was termed Excitability-Inducing Material (EIM).

The plot shown in Figure 7.2 can be quantitatively analyzed by assuming, first, that a single protein is responsible

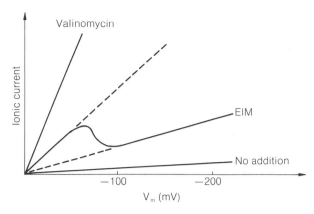

FIGURE 7.2. Effects of valinomycin and EIM on the ionic conductance across an artificial lipid bilayer. An aqueous electrolyte solution containing 0.1M KCl and an appropriate buffer was placed in two compartments, separated by a small orifice. A lipid bilayer was then constructed across the orifice, and the membrane conductance was measured as a function of the applied electric field. The experiment was performed in the presence or absence of an ion-conducting agent.

for the enhanced ionic conductance; second, that this protein can exist in either of two conformations; third, that these two conformations conduct ions at different efficiencies; and fourth, that the relative thermodynamic stabilities of the two protein conformations are influenced by the transmembrane potential. An increase in the potential would tend to favor its low conductance conformation. This "two-state hypothesis," therefore, suggested that the following equilibrium situation existed:

$$S_1 \underset{\longleftarrow}{\overset{K}{\longrightarrow}} S_2$$

(high conductance state)   *(low conductance state)*

$$K = S_2/S_1$$

where the value of the equilibrium constant, K, which determines the relative concentrations of the two conformational states of EIM in the membrane, is determined by the membrane potential.

   If this hypothesis is correct, it should be possible to insert a single molecule of EIM into a synthetic membrane and observe a change in conductance as a function of time. When the protein "flips" from its low conductance state (S_2) to its high conductance state (S_1), an increase in the transmembrane current should be observed. Because of the high resistance of artificial membranes to alkali metal ions and chloride, the proposed experiment was feasible. The results are shown in Figure 7.3. With a single EIM molecule

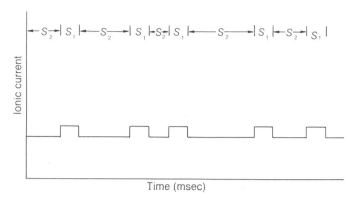

FIGURE 7.3. Transmembrane conductance of a synthetic membrane into which a single molecule of EIM is presumed to have been inserted. The electric potential across the membrane was held constant.

in the membrane,* the magnitude of the conductance at any one instant in time was of either of two values, and these two values were independent of the membrane potential. When the potential was low, EIM existed in a high conductance state (S_1) most of the time, but if the potential increased, the molecule could be made to exist predominantly in its low conductance state (S_2). These results confirm the two-state hypothesis and suggest a mechanism by which the activities of potential-sensitive and chemical-sensitive ionic channels are regulated.

Ion-Conducting Channels in Excitable Membranes

To explain their bioelectric properties, it is postulated that excitable membranes contain ion-specific pores that function as EIM does in an artificial membrane. The Na^+ and K^+ gates are two presumed examples of voltage-dependent ion channels in the nerve and muscle cell. Additionally, channel proteins, which are subject to regulation by chemical agents, must be postulated to account for neurotransmitter sensitivity. A probable example of a chemically sensitive ion pore is the acetylcholine receptor, an integral membrane protein complex found in the postsynaptic membranes of nerves and muscles. It is hypothesized that the binding of acetylcholine to its receptor protein opens an ionic gate that increases the permeability of the membrane to such small ions as K^+, Na^+, and Ca^{2+}. Under normal conditions, the ionic fluxes that result would lead to a reduction in the resting membrane potential and the initiation of an electric impulse (an action potential).

The acetylcholine receptor complex can be localized and identified on the basis of its high affinity association with

* Possibly a multimeric complex.

certain snake venom toxins, such as α-bungarotoxin, an inhibitor of receptor function. Employing such toxins, acetylcholine receptors can be solubilized and purified from several sources. The receptor protein from the electric organ of *Torpedo californica* is a polypeptide chain with a molecular weight of about 40,000*; it possesses both neurotransmitter and toxin-binding activities. The purified material can be inserted into a vesicular membrane system in a fashion which renders all of the toxin-binding sites exposed to the external surface of the vesicles. This system has been employed to demonstrate that the receptor can catalyze the chemically responsive transport of ions. Lipid bilayer vesicles containing the acetylcholine receptor protein were loaded with radioactive Na^+, and $^{22}Na^+$ efflux was measured in the presence and absence of the neurotransmitter. Acetylcholine was found to specifically enhance the rate of $^{22}Na^+$ efflux. Moreover, this effect was blocked by α-bungarotoxin, which functions as an antagonist of acetylcholine receptor function as noted. The purified receptor macromolecule alone evidently contains the molecular elements necessary for ion translocation and neurotransmitter-induced postsynaptic depolarization.

Preliminary progress has been made in characterizing the Na^+ channel found in nerve membranes. By the use of tetrodotoxin, a neurotoxin from the Japanese pufferfish, it is possible to selectively block the Na^+ conductance and the action potential without altering the K^+ conductance or the resting membrane potential. This observation has led to the suggestion that the Na^+ and K^+ channels are distinct, a conclusion also supported by electrophysiologic studies. Although a substance containing tetrodotoxin-binding activity has been extracted from nerve membranes with detergents, the material obtained was of a high molecular weight and has not been extensively purified.

It is clear from this brief discussion that our understanding of the molecular basis of membrane excitability is still fragmentary. The availability of a microbial system that exhibits excitable properties and is subject to genetic manipulation would be of obvious benefit in the elucidation of the mechanism of the gated response.

Bioelectric Control of Ciliary Activity in Paramecium

Paramecium is an elongated ciliated protozoan with bipolar symmetry. The cell anterior is characterized by a gullet for ingestion, and an anal pore is found posteriorly. Underneath the plasma membrane is a cytoskeleton, which confers upon the cell its shape and serves as an anchor for the hundreds of cilia that cover and are continuous with the

---

* The active membrane complex possibly exists as a tetramer of this polypeptide chain.

cell surface. *Paramecium* is capable of forward and backward motion because it can coordinately orient its cilia. Normally, the organism swims in the forward direction at a constant rate. If the anterior end of the cell comes into contact with an object, *Paramecium* will undergo reversal. The organism will orient its cilia anteriorly, and swim backwards. Seconds after reversal is initiated, however, forward motion is usually restored in a new and randomly chosen direction. By contrast, if an object touches the posterior end of *Paramecium*, it exhibits accelerated forward swimming.

The mechanisms of ciliary reversal and of accelerated motility have been extensively studied. As a result of these studies, it has been proposed that ciliary reversal is caused by Ca^{2+} influx and that accelerated forward swimming is caused by K^+ efflux. Normally the concetration of Ca^{2+} in the extracellular medium is high (10^{-4} to $10^{-3}M$) relative to its concentration in the cytoplasm (10^{-8} to $10^{-7}M$). Just the reverse is true for K^+. The intracellular K^+ concentration has been estimated to be about 40 mM, whereas the extracellular concentration is normally one to two orders of magnitude lower.

Electrophysiologic measurements have shown that an *anterior* stimulus is a *depolarizing* stimulus, which is propagated over the entire cell surface. Chemical stimuli can also cause depolarization and ciliary reversal. The stimulus apparently causes the Ca^{2+} channels to open so that Ca^{2+} flows into the cytoplasm. Although the amount of Ca^{2+} that enters the cell may be small, it is sufficient to appreciably enhance the local Ca^{2+} concentration in the cilia. By contrast, a *posterior* stimulus has been shown to be a *hyperpolarizing* stimulus, which spreads over the entire cell surface. This stimulus is thought to open the K^+ gates so that K^+ flows out of the cell into the medium. Depressed K^+ concentrations at the inner membrane surface may produce acceleration.

Several observations substantiate these hypotheses. An involvement of Ca^{2+} in reversal has been suggested by the finding that the magnitude of the depolarization that resulted from anterior stimulation was enhanced when the extracellular Ca^{2+} concentration was increased. Conversely, decreasing the external Ca^{2+} concentration reduced the depolarization until at $10^{-7}M$ Ca^{2+}, no signal, and no ciliary reversal, was observed. Changes in the extracellular K^+ concentration had no effect on reversal, but strongly influenced both hyperpolarization and the enhanced rate of ciliary beating elicited by a posterior stimulus. An increase in the K^+ concentration in the medium decreased the magnitude of the hyperpolarization, as would be expected, if enhanced membrane K^+ conductance were responsible for the signal. It can, therefore, be inferred that depolarization was due to Ca^{2+} influx, whereas hyperpolarization was due to K^+ efflux.

In another series of experiments, *Paramecium* cells were killed and extracted with the detergent, Triton X-100. This detergent treatment partially disrupted the cell membrane so that it was freely permeable to all ions. Under these conditions, ciliary orientation was subject to control by the extracellular Ca^{2+} concentration. If the *Paramecium* skeletons were suspended in a medium containing 10^{-8} to 10^{-7}M Ca^{2+}, the cilia were oriented posteriorly; but when the Ca^{2+} concentration was increased above 10^{-6}M, the cilia reversed direction. If, subsequently, ATP, $MgCl_2$, and KCl were added, the skeletons actually began to swim in the direction dictated by the concentration of Ca^{2+} in the medium!

Although the details of the molecular mechanisms controlling the gated ionic responses are not known, mutants that are defective in their bioelectric control processes have been isolated. Thus, one mutant type (called *pawn*) had lost the ability to undergo reversal. An anterior stimulus had no depolarizing effect on this mutant, which suggested that the Ca^{2+} channels were fixed in the closed position. A second mutant (called *paranoiac*) exhibited continuous ciliary reversal, lasting as long as a minute, in the presence of Na^+. Possibly, the genetic defect in this mutant facilitated the opening of the Ca^{2+} channels, so that Ca^{2+} conductance was enhanced. Finally, a third mutant (called *fast-2*) showed the opposite behavior from the *paranoiac* mutant: when transferred to Na^+-containing solutions, it exhibited no ciliary reversal for 5 or 10 min; instead, the membrane hyperpolarized, and the cell swam forward with increased velocity. Possibly, the genetic defect in this mutant affected the K^+ gate so that the K^+ permeability was abnormally high in the presence of Na^+. The behavior of the normal (wild-type) *parmecium* in Na^+ medium was intermediate between that of *paranoiac* and *fast-2*: It showed frequent, but not continuous avoiding reactions. Biochemical analyses of these genetically altered strains may lead to the identification of the proteins controlling transmembrane ion permeability and to the mechanism of the gated response.

Photoreception in the Rod Cell of the Mammalian Eye

Due to the great sensitivity of the retina's photoreceptor cells, the mammalian eye can function in exceptionally low light intensity. High sensitivity to light is a result of the high photoreceptor concentration, which allows the probability that an incident photon will be absorbed to approach unity. Moreover, a single photon absorbed anywhere in the outer segment has a high probability of eliciting a response

and generating a signal in the associated nerve cell. Since
the receptor molecules are at one end of the cell, and the
synapse is at the other, each cell must have an internal system of amplification and communication, allowing the signal to be transmitted from one cell to the other. Structural
features of the receptor cell and the mechanism governing
the response must account for these features.

A schematic drawing of the rod cell is shown in Figure
7.4. The cell consists of two parts: the inner segment that
contains the cellular metabolic machinery and the outer
segment that contains the differentiated disc membranes
where the photosensitive rhodopsin pigment is localized.
At the base of the inner segment is a synapse through
which an impulse can be propagated to a connecting neuron that leads to the optic nerve. The inner and outer segments are joined by a thin *connecting cilium*. Because of
the fragility of this structure, the outer segments can be
isolated free of contaminating materials after mild mechani-
cal agitation and sucrose density gradient centrifugation.
The disc membranes can then be isolated after osmotic
lysis.

Compositional analyses of rod outer segment membranes
show that, on a weight basis, about 50% of the membrane
consists of polar lipids; the other 50% is protein. Of the
membrane protein, 80 to 90% is rhodopsin, an integral protein that serves as the photoreceptor. Other proteins, in-

FIGURE 7.4. Schematic diagram of the rod
photoreceptor cell.

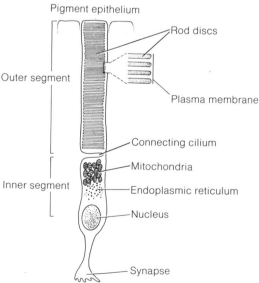

cluding ion-translocating ATPases and enzymes involved in cyclic AMP metabolism, are also thought to be present.

Rhodopsin is very likely an elongated protein about 40,000 daltons in size. It is thought to be exposed to the cytoplasmic surface of the disc membrane and may penetrate the entire structure. Although it is specifically oriented across the membrane, it can diffuse laterally through the lipid bilayer and rotate freely. The light-sensitive chromophore, 11-*cis*-retinal, is bound to the protein through a protonated Schiff's base and is oriented parallel to the plane of the disc membrane. The structure of the chromophore is shown in Figure 7.5.

The visual response involves several steps: photon absorption, signal generation, transmission and amplification, and conversion of the signal into synaptic activity. Light causes photoisomerization, the conversion of the chromophore, 11-*cis*-12s-retinal, to the all-*trans* configuration, and this reaction is followed by a series of protein conformational changes in which each conformation possesses distinct absorption properties and thermal stability. Exactly which of these photointermediates elicits the cellular response to photon absorption is still unclear. Further, the detailed mechanism by which the photoresponse is generated is not established. The mechanism proposed below takes into account available phenomenologic observations and is an attempt to describe the photoresponse in molecular terms.

FIGURE 7.5. (A) Structure of retinal in the all-*trans* configuration. (B) Structure of 11-*cis* retinal.

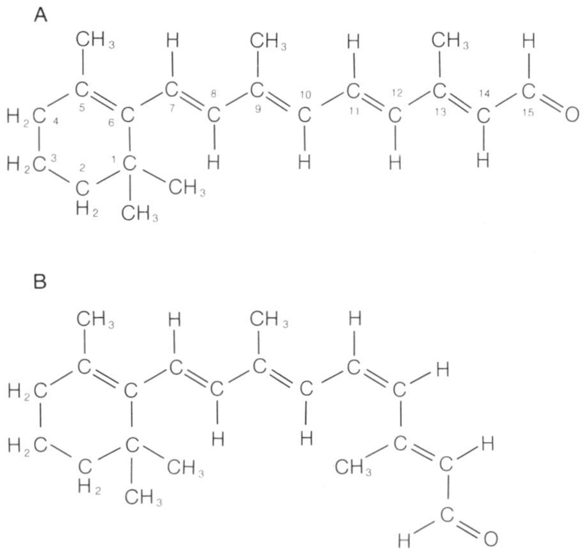

In the resting state, Na^+ flows into the outer segment of the rod cell via Na^+ channels in the outer segment plasma membrane. It is pumped out of the cell by the Na^+, K^+-translocating ATPase, localized in the plasma membrane of the inner segment. The rate of ion flow is sufficient to allow the intracellular Na^+ to exchange with the medium in about 45 min. Since the rates of Na^+ flow in each direction are equal, the membrane potential is maintained at an equilibrium value.

It has been proposed that Ca^{2+} is sequestered within the rod outer segment discs. The primary response to light may be to open ionic channels, which specifically allow Ca^{2+} to flow out of the disc sacs into the outer segment cytoplasm and, thus, to control the activity of the Na^+ gates in the plasma membrane: An increase in cytoplasmic Ca^{2+} may close the Na^+ gates, thereby decreasing the flow of Na^+ into the cell. Inhibition of the flow of positive charge into the cell, without a corresponding inhibitory effect on the Na^+ pump, should produce hyperpolarization; inside, the membrane potential should become more negative than during the resting state. Finally, hyperpolarization is transmitted throughout the cell to the synaptic region, where the electric signal is converted into synaptic activity.

Most available evidence appears to support this hypothesis. Extracellular Na^+ is required for ion flow in the resting state and for the photoresponse. Both processes can be blocked with metabolic inhibitors or ouabain. Involvement of Ca^{2+} is suggested by the fact that the amplitude of hyperpolarization is influenced by the Ca^{2+} concentration in the bathing solution. Moreover, a Ca^{2+} ionophore, which penetrates the membrane and allows Ca^{2+} to enter the cytoplasm, mimics the effect of light. The hypothesis also explains the amazing fact that a single photon elicits a current equal to about a million Na^+ ions after a lag period of 1 to 2 msec following illumination. By invoking Ca^{2+} as a "messenger," it is possible to account for this large amplification, and diffusion of Ca^{2+} through the cytoplasm to the plasma membrane accounts for the lag period.

Bacterial Photoreception and Transmission

Halobacterium halobium is a motile halophilic bacterium, which naturally inhabits salt flats at the edges of tropical seas. It is frequently exposed to environments of low oxygen tension and high light intensity and, therefore, should derive particular benefit from a mechanism that facilitates the conversion of solar energy into bioenergy. The protein molecule that accomplishes this feat is bacteriorhodopsin. Properties of this molecule are summarized and compared with those of mammalian rhodopsin in Table 7.1.

When *H. halobium* is grown aerobically, in the absence

Table 7.1   Properties of Mammalian and Bacterial
Rhodopsin

Property	Mammalian rhodopsin	Bacterial rhodopsin
Molecular weight	40,000	26,000
Globular elongated protein which penetrates the membrane	+	+
Retinal as chromophore	+	+
Chomophore bound to a lysyl residue in Schiff's base linkage	+	+
Light produces protein photointermediates	+	+

of light, little bacteriorhodopsin is synthesized. But, oxygen depletion and exposure to the light result in a 100-fold enhancement of net bacteriorhodopsin production. The molecule seems to be the sole protein constituent of purple membrane patches which are embedded in the plasma membrane of the cell; bacteriorhodopsin molecules apparently associate with each other and with phospholipids (25% by weight) to form membrane segments that are stable relative to the rest of the plasma membrane. Consequently, these photosensitive purple membrane patches can be isolated free of other cell constituents.

Physical analyses of the isolated membrane segments suggest that the protein is essentially immobile, since it shows no lateral or rotational motion. Due to specific bacteriorhodopsin associations, the molecules associate in a planar hexagonal lattice.

Like mammalian rhodopsin, the bacterial protein has an elongated globular structure that appears to extend through the membrane. The chromophore, all-*trans* retinal, forms a Schiff's base linkage to an ε-amino group of a lysine residue in the protein, reminiscent of the mammalian protein. Illumination initiates a series of cyclic protein conformational changes, which are recognized by changes in absorption. The cycle is complete within a few milliseconds. The protein photointermediates probably resemble the conformational states noted for mammalian rhodopsin. Still more significantly, illumination of a bacterial suspension also acidifies the medium: bacteriorhodopsin apparently catalyzes the electrogenic transport of protons across the membrane so that a pH gradient and a membrane potential (negative inside) are created. That this pumping activity is, in fact, due to the bacteriorhodopsin molecule was shown by inserting purple membrane patches in a synthetic lipid bilayer and demonstrating pumping activity. Estimation of the stoichiometry of the reaction, employing whole cells, suggests that one proton is transported per photon absorbed.

Several consequences of light absorption and the generation of proton and electric gradients have been noted. First, solar energy can be used to synthesize ATP by a non-photosynthetic mechanism in the absence of oxidative metabolism; second, light can be used to energize the active accumulation of solute molecules in the bacterial cell; third, the organism exhibits a phototactic response, indicating that the photoreception function of bacteriorhodopsin can influence the bacterial flagellum; and, finally, illumination of purple bacteria in the presence of oxygen has an inhibitory effect on respiration. It appears that throughout evolution the mammalian and bacterial proteins have retained similar structural and functional features, which allow them to catalyze light-triggered ion translocation. They have diverged only so that this function can be maximally utilized for the benefit of the organism. Whereas mammalian rhodopsin functions in the highly differentiated process of vision, bacteriorhodopsin has evolved into a multifunctional protein engaged in energy generation.

Selected References

Bean, R. C. "Protein-mediated mechanisms of variable ion conductance in thin lipid membranes," in *Membranes, A Series of Advances 2, Lipid Bilayers and Antibiotics* (G. Eisenman, ed.). Marcel Dekker, Inc., New York, 1973, p. 409.

Daemen, F. J. M. Vertebrate rod outer segment membranes. *Biochim. Biophys. Acta, 300*:255 (1973).

Eckert, R. Bioelectric control of ciliary activity. *Science, 176*:473 (1972).

Fambrough, D., H. C. Hartzell, J. E. Rash, and A. K. Ritchie. Receptor properties of developing muscle. *Ann. N.Y. Acad. Sci., 228*:47 (1974).

Fuortes, M. G. F. (ed.). "Physiology of photoreceptor organs," in *Handbook of Sensory Physiology, Vol. VII.* Springer-Verlag, Berlin, 1972.

Hagins, W. A. "The visual process: Excitatory mechanisms in the primary receptor cells," in *Annual Review of Biophysics and Bioengineering, Vol. 1.* Annual Reviews, Inc., Palo Alto, Cal., 1972, p. 131.

Katz, B. *Nerve, Muscle, and Synapse.* McGraw-Hill, Inc., New York, 1966.

Kung, C., S. Y. Chang, Y. Satow, J. Von Houtan, and H. Hansma. Genetic dissection of behavior in *Paramecium. Science, 188*:898 (1975).

Michaelson, D. M., and M. A. Raftery. Purified acetylcholine receptor: Its reconstitution to a chemically excitable membrane. *Proc. National Acad. Sci. USA, 71*:4768 (1974).

Mueller, P. and D. O. Rudin. "Translocators in bimolecular lipid membranes," in *Current Topics in Bioenergetics, Vol. 3* (D. R. Sanadi, ed.). Academic Press, New York, 1969, p. 157.

Oesterhelt, D. and W. Stoeckenius. Functions of a new photoreceptor membrane. *Proc. Nat. Acad. Sci. 70:*2853 (1973).

Racker, E. and W. Stoeckenius. Reconstitution of purple membrane vesticles calalyzing light-driven proton uptake and adenosine triphosphate formation. *J. Biol. Chem., 249:*662 (1974).

Stoeckenius, W. and R. H. Lozier. Light energy conversion in *Halobacterium halobium. J. Supramolecular Structure, 2:*769 (1974).

Wald, G. Molecular basis of visual excitation. *Science, 162:*230 (1968).

8 Hormonal Regulation of Cellular Metabolism

To me life consists simply in this, in the fluctuation between
two poles, in the hither and thither between the two
foundation pillars of the world.

Hermann Hesse

Several classes of animal hormones are known, all of which
exert control over intracellular metabolism. Steroid hormones, for example, are thought to penetrate the plasma
membrane, bind to cytoplasmic receptors, and travel to the
nucleus where the hormone receptor complex influences
gene expression. By contrast, prostaglandins, catecholamines,
and peptide hormones probably exert their regulatory effects at the level of the plasma membrane. Specific hormone receptor proteins have been postulated to account
for the action of the latter compounds on the activities of
membrane-associated enzyme systems that catalyze solute
transport and the synthesis of cyclic nucleotides. Prostaglandins are water-insoluble compounds, which probably
intercalate into the phopholipid matrix of the membrane
before binding to a hydrophobic receptor site within the
membrane. Some evidence exists to suggest that prosta-
glandin receptors (which have not been isolated or characterized biochemically) may mediate the action of other
hormones and thus function as transducer elements in the
membrane. Catecholamines (epinephrine, norepinephrine)
are simple derivatives of the amino acid, tyrosine, and can
penetrate most biologic membranes via specific transport
systems. The characterization of the membrane receptors
that are presumed to mediate the action of these hormones
is also in its infancy. Considerably more information is
available concerning the interactions responsible for the

regulatory effects of peptide hormones (insulin, glucagon, growth hormone, etc.). These macromolecules apparently bind to integral receptor proteins on the external surface of the plasma membrane, exerting their effects without entering the cytoplasm. In this chapter, the nature of peptide hormone receptors and possible mechanisms of their action will be considered. An analogy will be drawn between hormonal control of cellular metabolism in animal cells and the regulation of membrane-associated enzyme systems in bacteria. It will be seen that, in these evolutionarily divergent organisms, the activities of adenylate cyclase and of nutrient transport systems are subject to regulation. In both biological systems, the modulation of intracellular cyclic AMP levels allows fine control over cellular metabolic rates. But the mechanisms by which metabolic control is achieved differ: In the animal cell the *activities* of catabolic enzymes are subject to direct regulation, whereas in the bacterial cell the *synthesis* of corresponding enzymes seems to be a primary site for control. It would appear that differing degrees of biological complexity have required the evolution of divergent molecular mechanisms for regulating cellular metabolism although a common function may have been retained.

The Insulin Receptor Insulin and other peptide hormones appear to influence the activities of several membrane-associated enzyme systems. One of these is adenylate cyclase, the enzyme that catalyzes the synthesis of cyclic AMP from ATP. Cyclic AMP is frequently thought of as a cytoplasmic messenger of hormone action. The binding of a hormone to its receptor in the plasma membrane may either increase or decrease the activity of adenylate cyclase, depending on the hormone, causing the intracellular level of the cyclic nucleotide to rise or fall, respectively. Intracellular cyclic AMP, in turn, controls the activities of protein kinases, which phosphorylate a variety of metabolic enzymes. The phosphorylation of such an enzyme may either enhance or depress its catalytic activity, depending on the enzyme. Hence, intracellular cyclic AMP levels that reflect hormone-receptor interactions on the external surface of the plasma membrane control the rates of cellular metabolism. This complex process provides one mechanism of endocrine control in a multicellular organism. Since hormones are produced in specialized glands, distant from their target organ, hormonal regulation allows for cellular communication at a distance.

Insulin binds to a limited number of sites in the plasma membrane of an insulin-responsive cell. There are about ten exposed binding sites/μ^2 on the fat cell membrane surface (about 10,000 sites/cell). This is a very low number when it is compared, for example, with the surface density of the acetylcholine receptor, which exhibits 100,000 sites/μ^2 in the postsynaptic membrane of the neuromuscular junction. Association between hormone and receptor occurs with high affinity ($K_d = 10^{-10}$M) and is highly specific. Although insulin preparations from various mammalian sources competitively inhibit binding of radioactive insulin to the cell surface, binding is not prevented by a number of structurally unrelated proteins. Moreover, proinsulin, the precursor polypeptide from which insulin is derived by proteolysis, binds to the receptor with an affinity which is 20-fold lower than that of insulin. That insulin-binding sites are asymmetrically distributed in the membrane has also been established; less than 2% of the binding sites are found on the cytoplasmic side of the bilayer. This observation implies an asymmetric biogenic insertion mechanism and, together with other observations, eliminates the possibility that the hormone-receptor complex must "flip" from one side of the bilayer to the other to exert its biological effect. The binding of a peptide hormone to the external surface of the membrane must be sufficient to elicit a response. These observations are consistent with the fluid mosaic model of membranes, in which only lateral diffusion, but not transverse movement of proteins, can occur to a significant degree.

Studies on insulin receptor function have provided convincing evidence that the receptor is a glycoprotein in which the sugar residues are required both for hormone binding and for transmission of the "signal" to a responsive enzyme, such as adenylate cyclase. If the cells are treated with sialidase, an enzyme that cleaves the monosaccharide, sialic acid, from terminal positions in the carbohydrate chains of glycoproteins, it is found that the modified receptor retains the capacity to bind insulin with high affinity but that insulin binding no longer elicits a biological response. If the sialidase-treated preparation is exposed to another glycosidase, β-galactosidase, which hydrolyzes terminal galactose residues from the carbohydrate chains, then binding activity is also lost. This binding capacity is *not* lost if cells not previously exposed to sialidase are treated with β-galactosidase, suggesting that the galactosyl residues are internal and that the sialic acid residues are in terminal positions linked to the galactosyl residues. These observations appear to establish a role for the carbohydrate chains in receptor function, but whether they participate directly in the binding reaction or are merely required to maintain the normal stability of an appropriate conformation of the polypeptide chain has not been determined.

An integral protein with insulin-binding activity has been extracted from fat and liver cell membranes with the non-ionic detergent, Triton X-100, and has been purified more than 100,000-fold. The solubilized complex has an approximate molecular weight of 300,000 and binds insulin with the same affinity as the membrane-associated receptor. Although the receptor has not yet been obtained in homogeneous form, its glycoprotein nature appears to be established, since proteolytic enzymes and glycosidases both abolish its ability to bind radioactive insulin. It is not yet known if the large insulin receptor complex possesses enzymatic activities associated with receptor function.

In addition to adenylate cyclase, the binding of insulin to its receptor appears to influence the activities of guanylate cyclase (which catalyzes the synthesis of cyclic GMP) and the glucose transport system. These catalytic systems may be subject to direct regulation by the insulin receptor, or they may be secondarily controlled.

Possible Mechanisms of Hormone Regulation

In the fat cell, adenylate cyclase and other enzyme systems are sensitive to regulation by several hormones in addition to insulin. For example, glucagon and epinephrine stimulate adenylate cyclase activity after binding to their respective membrane-associated receptors. It appears that several membrane-associated enzyme systems are subject to regulation by a variety of hormone-receptor complexes. The mechanism by which the binding of a hormone to its receptor protein regulates adenylate cyclase and other membrane-associated proteins is not known, but several possibilities can be envisaged. In Figure 8.1, a mechanism is suggested in which the receptor protein possesses a site that can bind specifically to an allosteric regulatory site of each of the enzymes subject to regulation. As depicted in Figure 8.2, the assumption is made that the membrane receptor can exist in either of two conformations. The $\boxed{R_I}$ conformation possesses an intramembranous, high affinity binding site specific for the allosteric regulatory site of the enzyme, but the (R_I) conformation lacks this site. In the absence of the hormone, the (R_I) conformation must predominate, but binding of hormone to its receptor shifts the equilibrium

FIGURE 8.1. A possible mechanism by which any of a variety of hormones [insulin (I), glucagon (G), or catecholamine (C)] may bind to their respective receptors (R) to influence the activities of various membrane enzymes (E₁), (E₂), and (E₃). The receptors are assumed to be exposed to the external surface of the plasma membrane, whereas the enzymes are localized on the cytoplasmic surface. The model proposes a direct interaction between hormone receptors and the enzymes subject to control.

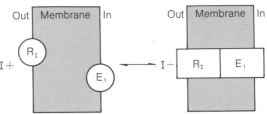

FIGURE 8.2. Direct interaction of a hormone receptor
with an enzyme in the membrane. Both the receptor
(R_I) and the enzyme (E_1) are presumed to be
capable of lateral diffusion through the plane of the
membrane. The binding of the hormone (I) to its
receptor (R_I) converts the receptor to a
conformation which binds E_1 with high affinity. The
binding of the hormone–receptor complex to a site
on E_1 is presumed to alter the catalytic activity
of the latter.

toward $\boxed{R_I}$. An interaction of $\boxed{R_I}$ with the allosteric regulatory site of the enzyme will either enhance or depress its activity, depending on the enzyme and the receptor involved. This model assumes a fluid membrane in which the integral membrane constituents can diffuse laterally.

The second model (Figure 8.3) suggests that the hormone receptor does not bind directly to the regulated enzyme but influences a "transducer" element in the membrane. This transducer may secondarily control the activities of membrane enzymes; it could, for example, be a protein that specifically binds to those enzymes subject to control, or it may catalyze a reversible reaction, which influences the enzyme activities. Relevant to this last possibility is the suggestion that adenylate cyclase (or a membrane protein that influences its activity) can exist in both the free and phosphorylated state. According to this hypothesis, the dephosphorylated form of the adenylate cyclase complex possesses maximal activity, and phosphorylation lowers its activity. Possibly, the binding of a hormone to its receptor directly influences the activity of a membrane-associated protein kinase, which catalyzes the phosphorylation of hormone-sensitive enzyme complexes.

At present, insufficient information is available to allow us to choose between the hypotheses. The resolution of this problem will probably require the isolation of all components of a hormone-responsive regulatory system and reconstitution of the system in a synthetic bilayer membrane. But, pertinent information might be forthcoming from a consideration of metabolic regulation in microorganisms.

FIGURE 8.3.

Indirect interaction of receptors and enzymes involving a
transducer element (T). The scheme does not distinguish a
mechanism in which the transducer plays a structural role
from one in which it functions catalytically (see text).
Abbreviations are as in Figure 8.1.

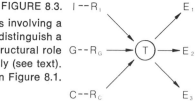

Regulation of Bacterial Metabolism

The phenomenon of catabolite repression of enzyme synthesis in bacteria has been recognized for decades. When
a sugar substrate of the phosphoenolpyruvate:sugar phosphotransferase system (PTS) is added to the medium of a
bacterial suspension utilizing a sugar, such as lactose, synthesis of the lactose catabolic enzyme system (including the
lactose transport protein and β-galactosidase) is inhibited.
To comprehend the basis for enzyme repression, the
molecular details of the process by which the lactose-
specific genes are transcribed to messenger RNA (mRNA)
must be understood.

Transcription of the lactose (lac) operon is under the
control of the lac repressor protein, which blocks transcription, and a cyclic AMP receptor protein (CR protein), which,
in the presence of cyclic AMP, promotes transcription (Figure 8.4). For transcription to occur, the repressor must be
removed from the operator region of the lac operon, and
the cyclic AMP–CR protein complex must be bound to the
promoter region of the operon. Dissociation of the repres-

FIGURE 8.4

Proposed scheme for the regulation
of the transcription of the lactose
operon in E. coli. The RNA
polymerase binds to a specific site
in the promoter region of the lactose
operon (lacP). It can transcribe the
DNA sequence of the structural
genes of the operon into
messenger RNA only if the cyclic
AMP-CR protein complex is bound
to the promoter region of the operon
(positive control) and if the
repressor is not bound to
the operator region (lacO) (negative
control). The scheme illustrates dual
control of transcription by two small
cytoplasmic molecules, cyclic AMP
and the inducer (I). The first
structural gene in the lac operon,
lacZ, codes for β-galactosidase.

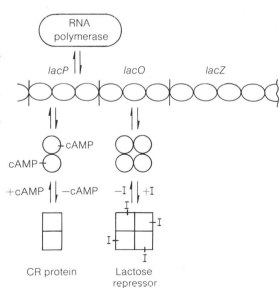

sor from the operator occurs whenever the inducer concentration in the cytoplasm is sufficiently high because the binding of inducer to the repressor converts the protein to a conformation that has a very low affinity for the DNA. In contrast, association of the CR protein with the promoter region of the *lac* operon *requires* the presence of cyclic AMP. Thus, the rate of transcription of the lactose genes will be depressed when the intracellular concentration of either the inducer or cyclic AMP is depressed. Reduction in the cellular levels of these compounds, brought about by addition of glucose or another sugar substrate of the PTS to the culture medium, should therefore repress β-galactosidase synthesis. The repression mechanism allows bacteria to select preferred sources of carbon for growth: glucose is utilized preferentially to lactose.

Several transport systems responsible for inducer uptake are subject to allosteric control by the PTS. As a consequence, addition of glucose or another sugar substrate of the PTS to a cell suspension inhibits the uptake of several carbohydrates (lactose, maltose, glycerol, melibiose, etc.) via their respective transport systems. Moreover, cellular cyclic AMP levels are depressed upon addition of glucose to the culture medium, first, because the activity of adenylate cyclase is inhibited, and, second, because the activity of a cyclic AMP transport system, which catalyzes excretion of the cyclic nucleotide from the cell, is stimulated. Regulation of adenylate cyclase and the carbohydrate transport systems appears to occur by a common mechanism, discussed below. Cyclic AMP excretion is considered in the next section.

Genetic analyses of transport regulation have suggested the scheme shown in Figure 8.5. A regulating sugar substrate of the PTS appears to interact with a specific membrane protein, the enzyme II responsible for the transport and phosphorylation of that sugar (see Chapter 5). Thus, mannitol can interact with the mannitol-specific enzyme II, glucose interacts with its enzyme II, and fructose interacts with the enzyme II specific for fructose. Any one of these interactions is presumed to lead to a change in a central regulatory protein, RPr, and, in some fashion, RPr regulates the activities of the permeases that transport lactose, maltose, melibiose, and glycerol. These interactions occur in a fashion such that the activities of these permeases are inhibited whenever a sugar substrate of the PTS is present in the culture medium. This scheme is based on the analyses of mutant strains that are resistant to transport regulation. In one such mutant that lacked the enzyme II for mannitol, all transport systems normally subject to regulation were found to be insensitive to inhibition by mannitol but were still sensitive to regulation by other sugar substrates of the

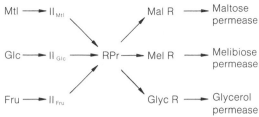

FIGURE 8.5. Proposed scheme for PTS-mediated transport regulation in bacteria. The scheme suggests that a sugar substrate of the PTS [mannitol (Mtl), glucose (Glc), fructose (Fru)] interacts with the enzyme II complex in the membrane specific for that particular sugar. The interaction of the sugar with the enzyme II causes a change in a central regulatory protein, RPr. Finally, RPr interacts with sugar-specific regulatory components, which control the activities of each of several different transport systems.

PTS. In an analogous mutant that lacked the enzyme II specific for fructose, all transport systems were resistant to inhibition by fructose, but mannitol and other sugar substrates of the PTS still exerted their normal regulatory effects. These characteristics implicated the enzymes II as sugar-specific components of the regulatory system. In a second class of mutants, all transport systems were insensitive to regulation by all sugar substrates of the PTS. These mutants appeared to lack, or be defective for, the central regulatory protein, RPr. Finally, in a third class of mutants, a single permease system was resistant to inhibition by all sugar substrates of the PTS, but the other transport systems retained normal sensitivity to inhibition. For example, the glycerol permease system was specifically rendered resistant to transport regulation by one such mutation, whereas another mutation abolished regulation of maltose permease activity. Genetic evidence led to the suggestion that this last class of mutants may have been defective for regulatory sites on the individual permease proteins.

Further genetic and physiologic studies showed that mutant strains with reduced cellular levels of enzyme I or HPr were *hypersensitive* to regulation; exceptionally low concentrations of a sugar substrate of the PTS were sufficient to inhibit transport activity. These observations provided evidence for the hypothetical mechanism shown in Figure 8.6. It is suggested that RPr can exist in either of two forms; one is phosphorylated, the other is not. Phosphorylation of RPr can only occur as a result of phosphate

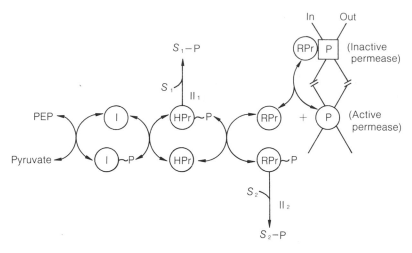

FIGURE 8.6. Proposed mechanism for PTS-mediated regulation of permease function in bacteria. The figure illustrates the phosphorylation of the central regulatory protein (RPr), as a result of sequential transfer of phosphate from phosphoenol-pyruvate (PEP) to enzyme I (I), to a heat-stable phosphate carrier protein (HPr), and, finally, to RPr. RPr may interact with an allosteric regulatory site on the permease to inhibit its activity. The phosphorylated form of RPr does not inhibit the permease. Addition of a sugar substrate of the PTS (S_1 or S_2) drains phosphate away from phospho-RPr, converting it to free RPr, which interacts with permease to inhibit its activity.

transfer from phospho-HPr. Only the nonphosphorylated RPr presumably can interact with the permeases to inhibit their activities. According to this proposal, the regulatory protein, RPr, is normally phosphorylated so that the transport systems are in an active form. But, addition of a sugar substrate of the PTS promotes dephosphorylation of this protein, converting it to the free form that can interact with the sensitive permeases to inhibit their activities. This proposal suggests that the catalytic activities of the PTS proteins are responsible for the regulation of transport activity.

Adenylate cyclase is apparently subject to regulation by a mechanism that also involves RPr and is analogous to the one proposed above for transport regulation. Thus, reduction in the cellular level of either enzyme I or HPr renders adenylate cyclase hypersensitive to regulation by a sugar substrate of the PTS, provided that the enzyme II complex specific for that sugar is functional. But, most mutants that lack RPr synthesize cyclic AMP at very low rates. This observation has led to the proposal that adenylate cyclase may be subject to allosteric regulation by phospho-RPr, which converts the enzyme from a conformation of low catalytic potential to one with enhanced activity. Thus, although the transport systems may be subject to *negative* control by RPr, adenylate cyclase may be subject to *positive* control by phospho-RPr. The net results, however, are similar: phosphorylation of RPr enhances the rates of inducer uptake and cyclic AMP synthesis. Addition of glucose, which causes dephosphorylation of RPr, reduces the rates of inducer uptake and of cyclic AMP synthesis. Coordinate regulation of cellular levels of inducer and cyclic AMP allows dual control over the rates of catabolic enzyme synthesis.

The relevance of these proposed mechanisms to hormonal regulation of membrane-associated enzyme systems in animal cells is not yet clear. For example, it is not known if the insulin receptor controls adenylate cyclase directly or by altering the rate of an enzymatic reaction. Nevertheless, in both biological systems, membrane-associated enzyme complexes involved in solute transport and cyclic AMP metabolism are the targets of regulatory control by external stimuli. Principles established with one system are likely to be applicable to the other.

Regulation of Cellular Cyclic AMP Levels

In the previous section, we noted that in several biological systems cyclic AMP plays an important role in the regulation of cellular metabolism. It also functions to regulate such differentiated processes as visual excitation, synaptic transmission, and cellular chemotaxis in multicellular organisms. Although the rates of cyclic AMP synthesis are clearly subject to regulation, variations in the intracellular concentrations of cyclic AMP reflect not only the synthetic rate but also the relative rates of cyclic AMP degradation and excretion.

The activity of mammalian cyclic AMP phosphodiesterase, the enzyme that hydrolyzes cyclic AMP to 5′-AMP, has been shown to be subject to regulation by a protein modulator or activator and Ca^{2+}. In the absence of the small heat-stable protein activator, the phosphodiesterase possesses low activity independent of Ca^{2+}. The activator, which specifically binds Ca^{2+} with high affinity, stimulates the catalytic activity of the phosphodiesterase at least tenfold, and this stimulation is strictly dependent on Ca^{2+}. It appears that the protein activator functions to render the activity of cyclic AMP phosphodiesterase sensitive to Ca^{2+} regulation. Thus, under physiologic conditions, cellular Ca^{2+} concentrations probably influence cyclic AMP levels by controlling the rate of cyclic AMP hydrolysis.

Cellular cyclic AMP concentrations can also be reduced by enhancing the rate of cyclic nucleotide excretion. Cyclic

AMP can be excreted into the extracellular fluid of both prokaryotic and eukaryotic organisms. In the bacterium, the transmembrane transport of the cyclic nucleotide presumably functions only to regulate cellular cyclic nucleotide concentrations, but, in other cell types, extracellular cyclic AMP may have a highly specific function. Thus, in the cellular slime molds, cyclic AMP in the culture medium functions as a chemotactic attractant that initiates the aggregation and differentiation of individual amoebae (see Chapter 6). In multicellular organisms, extracellular cyclic AMP may function, for example, in the regulation of neurobehavior.

Studies on cyclic AMP transport have shown that in both bacterial and mammalian cells efflux of cyclic AMP apparently occurs by energy-dependent processes. In the bacterium, an energized membrane state, possibly the transmembrane electric potential, provides the requisite energy for cyclic nucleotide transport (Chapter 5). In the animal cell, a chemical form of energy, such as ATP, may be involved. In both systems, rates of cyclic AMP transport are subject to regulation: in bacteria, the cellular energy level determines the rate of cyclic nucleotide efflux, whereas the same process in mammalian cells appears to be hormonally regulated by prostaglandins. From these considerations, it seems clear that cellular cyclic AMP levels are subject to regulation by several independently functioning systems. These systems may act coordinately to increase or decrease the rates of specific cell processes, or they may act in opposition to one another to permit fine control through different environmental and cellular conditions. The need for complex cellular regulatory mechanisms becomes apparent when it is realized that the unregulated activities of the cellular metabolic machinery would result in massive waste of available nutrients and energy, a condition inconsistent with life. Complex regulatory processes have created efficient biological machines with the potential to evolve toward increasing degrees of complexity.

Selected References

Constantopoulos, A. and V. A. Najjar. The activation of adenylate cyclase: The postulated presence of adenylate cyclase in a phospho (inhibited) form and a dephospho (activated) form. *Biochem. Biophys. Res. Commun., 53*:794 (1973).

Cuatrecasas. P. Insulin receptor of liver and fat cell membranes. *Fed. Proc., 32*:1838 (1973).

Davoren, P. R. and E. W. Sutherland. The effect of L- epinephrine and other agents on the synthesis and release of adenosine 3' 5' phosphate by whole pigeon erythrocytes. *J. Biol. Chem., 238*: 3009 (1963).

Doore, B. J., M. M. Bashor, N. Spitzer, R. C. Mawe, and M. H. Saier, Jr. Regulation of adenosine 3':5'-monophosphate efflux from rat glioma cells in culture. *J. Biol. Chem., 250*:4371 (1975).

Magasanik, B. "Glucose effects: Inducer exclusion and repression," in *The Lactose Operon* (J. R. Beckwith and D. Zipser, eds.). Cold Spring Harbor Laboratory, Cold Spring Harbor, New York, 1970, p. 189.

Makman, R. S. and E. W. Sutherland. Adenosine 3′ 5′ phosphate in *Escherichia coli. J. Biol. Chem., 240*:1309, (1965).

Pastan, I. and R. L. Perlman. Cyclic adenosine monophosphate in bacteria. *Science, 169*:339 (1970).

Robison, G. A., R. W. Butcher, and E. W. Sutherland. *Cyclic AMP.* Academic Press, New York, 1971.

Roseman, S. "A bacterial phosphotransferase system and its role in sugar transport," in *The Molecular Basis of Biological Transport* (J. F. Woessner, Jr., and F. Huijing, eds.). Academic Press, New York, 1972.

Saier, M. H., Jr. and B. U. Feucht. Coordinate regulation of adenylate cyclase and carbohydrate permeases by the phosphoenolpyruvate:sugar phosphotransferase system in *Salmonella typhimurium. J. Biol. Chem., 250*:7078 (1975).

Saier, M. H., Jr., and S. Roseman. Inducer exclusion and repression of enzyme synthesis in mutants of *Salmonella typhimurium* defective in enzyme I of the phosphoenolpyruvate: Sugar phosphotransferase system. *J. Biol. Chem., 247*:972 (1972).

Tao, T. S. and J. H. Wang. Mechanism of activation of a cyclic adenosine 3′ 5′-monophosphate phosphodiesterase from bovine heart by calcium ions. *J. Biol. Chem., 248*:5950 (1973).

9 Cell Recognition

And the end of all our exploring
Will be to arrive where we started
And know the place for the first time.

T. S. Eliot

Intercellular adhesions are important in a wide variety of organisms, from bacteria to man. Organelles of adhesion are thought to provide certain bacteria with devices for securing nutrients and facilitating genetic exchange. In higher organisms, adhesive forces are presumed to be of prime importance during embryonic development and in the maintenance of distinct structural features of tissues in the adult organism. Intercellular adhesion also facilitates such diverse processes as fertilization, nutrient absorption, and excretion; and the presence of macromolecular receptors on animal cell surfaces allows cellular infection by viral and bacterial agents to occur. Moreover, the normal adhesive properties of a cell may be lost or modified under abnormal conditions, such as when transformation to malignancy occurs.

Despite the importance of intercellular adhesion in numerous biological processes, experimental data defining the phenomenon in molecular terms are largely lacking. Only in a few instances have the cell surface constituents that mediate intercellular interactions been identified and isolated. The paucity of information has generated abundant hypotheses on the chemical nature of cellular adhesion. For example, one hypothesis originated with the observation that glycosyl transferases, which catalyze the transfer of sugars from sugar nucleotides to appropriate acceptor glycoproteins or glycolipids, are associated with the surface plasma membranes of certain animal cells. The suggestion was advanced that cell adhesion may result from interactions between the glycosyl transferases on one cell and the surface glycosyl acceptors of another cell. Since each glycosyl transferase exhibits specificity for its acceptor molecule, highly specific cell interactions would be possible. Moreover, the catalytic activity of the enzyme might induce changes in the surface carbohydrates of an adjacent cell, thereby influencing metabolic processes within that cell. Since this hypothesis can account for the high degree of specificity observed in cell recognition, and can explain intercellular modification, it is an appealing one. Moreover, as discussed below, there is evidence that catalytic proteins may function in specific types of adhesion. In the following paragraphs, we shall discuss a few investigations that have provided concrete information on the nature of the molecular constituents that mediate intercellular adhesion. We will consider homotypic adhesion between like cells, heterotypic adhesion between unlike cells, and the adsorption of bacterial viruses to their host cell surfaces.

Adhesion of Bacteriophage to Bacterial Hosts

The first step in viral infection is generally the adsorption of the virus to the host cell surface. Adsorption of a bacterial virus (a bacteriophage) to its host cell apparently depends on the presence of attachment sites on the surface of the bacterium and a complementary site on the phage, which serves as the recognition unit. Evidence for an involvement of complementary macromolecules in phage adsorption initially came from studies showing that phage could adhere to heat-killed bacteria, and X-ray–inactivated phage could still adsorb to bacteria. It is now known that phage receptors on the bacterial surface are usually specific proteins, lipopolysaccharides, or components of the peptidoglycan layer of the cell wall. Bacterial mutations, which render a host resistant to phage attack, frequently alter the cell surface macromolecule that functions as the receptor. Such mutations presumably destroy complementarity with the phage organelle of adhesion. Analogous phage mutants that adsorb to an *altered* cell surface receptor have been isolated. In these mutants, the phage organelle of attachment is presumably altered.

Phage with contractile tail sheaths (*E. coli* phage T2 and T4) initially attach to their bacterial hosts by means of long tail fibers, which recognize specific attachment sites in the lipopolysaccharide layer of the Gram-negative cell. Isolated lipopolysaccharide fragments retain the capacity to bind to the phage tail fibers, and specific sugar residues in the lipopolysaccharide, which must be of the proper anomeric configuration, are crucial to this process. Moreover, since binding occurs rapidly at 0°C, enzyme catalysis is probably not involved. The protein in the tail fiber that recognizes a

specific sequence of sugar residues in the bacterial lipopolysaccharide must be a complementary sugar-binding
protein.

Adsorption of phage with short noncontractile tails (Salmonella phage ε-15 and P22) to lipopolysaccharide possibly involves an endoglycosidase located in the phage tail.
Attachment of phage P22 or of isolated tail parts to Salmonella cells is followed by spontaneous detachment.
Concomitantly, the lipopolysaccharide receptor site is de-
stroyed, and a rhamnosyl α-(1,3)-galactosyl linkage within
the polymer is hydrolyzed. The enzyme exhibits specificity
toward the same molecular grouping that functions as the
receptor. Thus, an enzyme with known catalytic activity
conceivably functions in the recognition process, allowing
the virus to adsorb to an appropriate host cell.

Heterotypic
Adhesion
—Sexual
Agglutination
in Yeast

Yeast pass through a sexual cycle in which diploid cells
sporulate with the formation of dormant haploid spores.
Each such spore can give rise to haploid vegetative cells,
which reproduce by budding. Such a vegetative cell may
adhere to and eventually fuse with a cell of opposite mating type when contact is made. The surface components
responsible for agglutination of Hansenula wingei cells of
opposite mating type have been extensively investigated.
Glycoprotein molecules can be released from the surfaces
of these haploid cells after treatment with appropriate auto-
lytic enzymes. One such factor (5-factor, released from
5-cells) consists of glycoproteins of heterogeneous size. The
material is a mannan-protein complex, in which as much as
90% of the net weight of the complex may be carbohydrate. The carbohydrate, which is covalently bound to the
polypeptide chains, apparently plays a secondary role in
stabilizing the protein and does not participate directly in
the adhesion process.

Extensive structural studies have shown that the carbohydrate, 5-factor polypeptide backbone linkage is different
from that in the bulk of the cell wall mannan proteins. In
the agglutination factor, chains averaging eight mannosyl
residues are attached to about one-half of the amino acid
residues in the polypeptide chain. The glycosidic linkages
are to the hydroxyl groups of serine and threonine residues.
In contrast, much longer carbohydrate chains are found in
the bulk cell wall glycoproteins, and the linkages appear to
involve asparagine or glutamine residues. This difference,
which is illustrated in Figure 9.1, establishes that the biosynthetic origin of the sexual agglutination factor is different from that of bulk cell wall glycoproteins, even though
both are localized to the cell wall fraction.

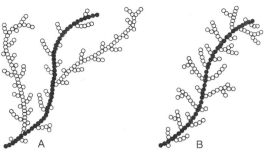

FIGURE 9.1. Postulated structures of (A) yeast cell wall
mannan glycoprotein in which most of the carbohydrate exists
as long polysaccharide chains, and (B) 5-agglutinin in which
most of the carbohydrate is in the form of short
oligosaccharides: mannose (○); amino acids (●).

After Yen and Ballou (1974).

Isolated 5-factor is apparently multivalent, since it can
agglutinate cells of the opposite mating type (21-cells). Its
specificity was demonstrated when it was found that 5-factor
would not agglutinate 5-cells or diploid strains, suggesting
that synthesis of its receptor is subject to repression by
genes carried by only one of the two haploid strains (by
5-cells). Synthesis of the factor also appears to be subject to
mating-type specific repressive control.

A surface mannan-protein, 21-factor, isolated from 21-
cells, has been implicated in the sexual agglutination process. Although it could bind to 5-factor and, thereby, inhibit
5-factor–promoted agglutination of 21-cells, it had no
agglutinating activity itself. It was therefore assumed to be
monovalent. As for the 5-factor, only the protein moiety of
21-factor appeared to be active in the agglutination process.
The 21-factor could not be isolated either from 5-cells or
from diploid strains, again suggesting that synthesis of the
protein is subject to mating-type specific regulation.

The properties of these cell aggregation factors indicate
that complementary macromolecules, located on the surfaces of cells of opposite mating type, are responsible for
the highly specific interactions of these cells. Although an
involvement of the protein moieties of these glycoproteins
has been suggested, the nature of their specific interactions
is still poorly understood.

**Homotypic
Adhesion
—Sponge
Aggregation**

Multicellular sponges of the *Microciona* genus consist of
loosely organized groups of differentiated cells. The cells
can be disaggregated as shown in Figure 9.2. The organism
is first cut into small pieces and then passed through a silk
cloth in calcium- and magnesium-free seawater. Subsequently, if the cells are removed from the medium by cen-

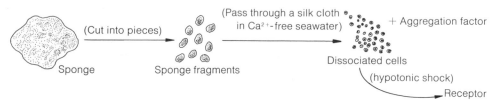

FIGURE 9.2. Procedure for dissociation of sponge cells and isolation of factors involved in intercellular adhesion.

trifugation, an *aggregation factor* can be isolated from the supernatant. In the absence of this factor, sponge cells will not reaggregate, but when the factor is added back to the cell suspension, agglutination immediately begins.

The aggregation factor has been purified and studied by physical techniques, including electron microscopy. It is a large acidic complex, having a molecular weight of two to five million. On a weight basis, the complex is about 50% amino acid residues, 50% sugar residues. The major sugars are hexosamines, glucose, mannose, and uronic acid. The fibrous glycoprotein complex is arranged in a unique sunburst configuration with an inner circle and about twelve radiating arms (Figure 9.3). Removal of Ca^{2+} results in dissociation of the arms into small subunits.

Aggregation factors have been isolated from several species. In each case, the molecules are species specific; they agglutinate the cells of the species from which the factor was isolated most effectively. Moreover, not all cells from a single species are aggregated with equal facility: one cell type from a *Microciona* species was found to form compact round aggregates, whereas another type from the same species formed loose aggregates. This difference might be due

FIGURE 9.3. Physical appearance of isolated aggregation factor from *Microciona* species.

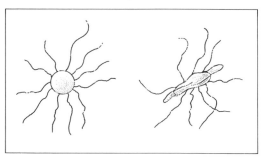

to differences in the number of receptors on the surface of the two cell types rather than a qualitative difference in the receptor molecule.

When dissociated sponge cells in 0.8M NaCl are subjected to hypotonic shock, the cells lose the capacity to associate in the presence of the aggregation factor. After removal of cells from the hypotonic shock fluid, one can demonstrate the presence in the supernatant of a factor that interferes with homotypic cell aggregation. Incubation of the hypotonic shock fluid with aggregation factor inhibits agglutination of an unshocked cell population. The material stimulates aggregation of *shocked* cells, however. The material responsible for these activities was found to be a small glycoprotein with a molecular weight of less than 100,000. The behavior quoted above implicates the molecule in the adhesion process and suggests that it functions as a *receptor* for the aggregation factor. But, because it could be released from the cell surface in a water-soluble form by hypotonic shock, it can be assumed that the receptor must be bound to the surface of the cells, possibly by an integral membrane protein.

Figure 9.4 illustrates how the receptor and aggregation factor may function in intercellular adhesion. The hypothetical molecular arrangement shown suggests that the receptor molecules possess specific binding sites for the radial fibers of the aggregation factor. The receptors, in turn, are associated with the cell surface, possibly by interaction with integral membrane constituents. A symmetric array of aggregation factors may typify homotypic adhesion.

Animal Cell Adhesion and Malignancy

The capacity of cells to adhere specifically to appropriate target cells is central to the process of tissue and organ formation in the developing embryo, and, for this reason, the mechanism of adhesion is of major interest to the biologist. Moreover, a pathologic condition might be expected to result if normal adhesion did not occur. The present section will be concerned with the presumed breakdown of cellular adhesion in malignancy and the consequences of neoplasia. A more general consideration of growth regulation will be undertaken in Chapter 10.

Development of a malignant tumor and onset of pathogenesis involves two distinct changes in cell behavior. The first alteration renders a normal cell insensitive to host growth regulatory signals and causes the cell to be permanently switched into an active growth cycle. If no further modification of cell behavior occurs, the tumor is categorized as benign (the tumor does not invade other tissue nor does it metastasize) and prospects for recovery are

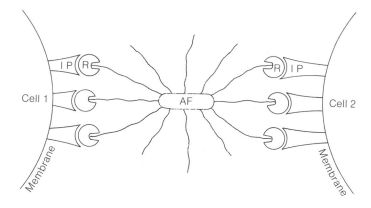

FIGURE 9.4. Proposed involvement of macromolecules in intercellular adhesion of sponge cells. Aggregation factor (AF); receptor (R); integral membrane protein (I P).

good if surgical removal is possible. But, subsequent to, or perhaps concomitant with loss of growth control, many tumor cells acquire the property of invasiveness. Dividing tumor cells invade surrounding tissue, disrupting normal organ architecture and inhibiting function. These tumor cells eventually invade the surrounding venules, break loose in the bloodstream or lymph system, lodge anew at distal sites, and form secondary tumors or metastases. Tumors that show histologic signs of invasiveness are categorized as malignant. Detection and removal of all tumor tissue are usually impossible, once metastasis has occurred.

Years ago, experiments were performed that indicated that a breakdown in cellular adhesion could be correlated with malignancy. Those tumors (of human lip and cervix origin) classified as malignant by histologic criteria were consistently less adhesive and easier to tease apart with dissecting needles than benign tumors of identical tissue. By inserting microwires into adjacent cells and measuring deflection of the wires as cells were pulled apart, it was possible to confirm quantitatively these qualitative observations: cells from normal tissue or benign tumors adhered to each other much more tenaciously than cells from malignant tumors. The results of this experiment clearly implicate defective intercellular adhesion as a primary cause of malignant tumor invasiveness.

Selected References

Crandall, M. A. and T. D. Brock. Molecular basis of mating in the yeast, *Hansenula wingei. Bacteriol. Rev., 32*:139 (1968).

Henkart, P., S. Humphreys, and T. Humphreys. Characterization of sponge agglutination factor. A unique proteoglycan complex. *Biochemistry, 12*:3045 (1973).

Kark, W. *A Synopsis of Cancer: Genesis and Biology.* Williams & Wilkinson Co., Baltimore, 1966.

Kuhns, W. J., G. Weinbaum, R. Turner, and M. M. Burger. Sponge aggregation: A model for studies on cell–cell interactions. *Ann. N.Y. Acad. Sci., 234*:58 (1974).

Lindberg, A. A. "Bacteriophage Receptors," in *Annual Review of Microbiology, Vol. 27.* Annual Reviews, Inc., Palo Alto, Cal., 1973, p. 205.

Roth, S. A molecular model for cell interactions. *Quarterly Review of Biology, 48*:541, (1973).

Yen, P. H. and C. E. Ballou. Partial characterization of the sexual agglutination factor from *Hansenula wingei* type 5 cells. *Biochemistry, 13*:2428 (1974).

10 Role of the Plasma Membrane In Growth Regulation and Neoplasia

> The cell is a wonderfully complex and precise little mechanism;
> disease is but disorder of this mechanism.
> The aim of medicine is to prevent or repair such disorders.
> The aim of biology is to understand the cellular machinery.
>
> A. Szent-Györgyi

In earlier chapters, we discussed the structures of biological membranes and the relationship of structure to basic membrane functions, such as compartmentation, transport, and molecular reception. We saw that, by and large, membranes in unicellular organisms perform the same functions as those in more complex biologic systems. In this chapter, regulatory interactions in multicellular organisms will be discussed, interactions that may prove to be far more complex than those occurring in unicellular organisms. We shall consider the role of the cytoplasmic membrane in mammalian cell growth regulation and carcinogenesis. In spite of the complexity of the problems at hand and the paucity of concrete information on this subject, the possible relevance of microbial studies to cell surface-mediated growth regulation will be discussed.

Growth of Normal and Transformed Animal Cells

In any study of cancerous growth, cell populations must be available that differ only with respect to oncogenicity (the ability to form a tumor in an appropriate host animal). "Normal" cells include fibroblast populations from primary tissue explants or established (immortal) fibroblast lines, such as the mouse line, 3T3. These cells are to the tissue culturist what *E. coli* is to the microbial physiologist. Primary fibroblasts and 3T3 cells are normal in that they do not form a tumor when injected into syngeneic host animals. But, infection of these cells by simian virus #40 (SV40) results in the integration of viral genetic material (DNA; MW, 400,000) into the host genome and the appearance of viral-specific antigens on the host cell surface. Instead of complete readthrough of the viral genetic material, which would result in virus production and cell lysis, only limited regions of the viral genome are expressed. Virus particles are not produced; instead, cellular morphology and growth behavior *in vitro* and *in vivo* are altered, and the cells are said to be transformed. These SV40-transformed 3T3 cells can form tumors in immuno-suppressed animals, whereas normal 3T3 cells cannot. Since the amount of viral information is minimal (sufficient to code for no more than ten polypeptide chains of average size), comparisons between oncogenic SV40 3T3 cells and normal cells become very meaningful.

In addition to SV40, mouse fibroblasts may be transformed by polyoma or murine sarcoma virus, by X-rays or by chemical carcinogens. The same kinds of modifications in cell morphology and growth behavior are observed regardless of the transforming agent, and these agents all induce changes in the cell surface antigens. Temperature-sensitive mutants of polyoma and Rous sarcoma virus have been isolated, which, in appropriate rat or chick host fibroblasts, give rise to a transformed phenotype at the permissive temperature but a normal phenotype at the restrictive temperature. The availability of these mutants has permitted elegant studies that allow unequivocal definition of the attributes of the transformed state.

Positive Growth Control   Tissue culture cells are unable to grow in the absence of a variety of factors in the culture medium. For example, optimal growth of mouse fibroblasts requires high concentrations of serum (about 10%) and a variety of sugars, amino acids, and vitamins. Individual serum factors may promote survival, growth, DNA synthesis, mitosis, and/or cell motility. Some serum growth factors have been purified to homogeneity and have been shown to be low molecular weight peptides structurally and functionally related to insulin. High concentrations of insulin mimic serum in promoting the growth of 3T3 cells, providing further evidence for a similarity between the modes of action of these peptides. The apparent structural and functional resemblance of these serum growth factors (somatomedins) to insulin has led to a wide acceptance of the notion that serum growth factors, like insulin, interact with external membrane receptors to regulate biochemical activities inside the cell.

Transformed cells typically require less serum for growth

than do normal cells. Whereas 10% fetal calf serum is required for optimal growth of 3T3 cells, SV40-transformed 3T3 cells grow satisfactorily in 1% fetal calf serum. In spite of this reduction in serum requirement, transformed cells can seldom grow in the total absence of serum; the serum requirement is reduced but not abolished. Possibly, transformation evokes changes in the plasma membrane, altering the quantity, distribution, or activity of surface receptors for serum growth factors.

In addition to serum factors, cells require numerous nutrients for growth. Conceivably, the rates of nutrient transport might limit and regulate cell growth. A conceptual problem in exploring this hypothesis involves distinguishing cause from effect. Do cells grow faster when nutrient uptake rates are enhanced, or do rapidly growing cells accumulate nutrients more efficiently? A considerable mass of experimental evidence now supports a growth regulatory role at the level of the transport systems. The rates of amino acid, nucleoside, and carbohydrate uptake into 3T3 cells are decreased when the cells attain confluency, and growth slows. For example, glucose transport is decreased tenfold when cells achieve confluency, and this decrease is also observed when transport of nonmetabolizable glucose analogues is followed. By contrast, uptake of glucose analogues into SV40 3T3 cells is not influenced by the confluent state, and growth of these cells continues after cell contacts have been made. When confluent 3T3 fibroblasts are shifted from a nongrowing to a growing state by addition of fresh serum or trypsin to the monolayer culture, the *earliest* detectable change in metabolic activity is a tenfold increase in the rate of glucose uptake. This change occurs 15 min after treatment, hours before a detectable increase in RNA, protein, or DNA synthesis is observed.

Finally, the observations cited above can be verified in temperature-sensitive transformed cells by monitoring transport at the permissive and restrictive temperatures. Aberrations in transport activity inevitably correlate with expression of the transformed state, which must reflect changes in the composition or structure of the plasma membrane.

Negative Growth Control

Cells in an adult animal are presumably bathed in interstitial fluid and low molecular weight nutrients, yet no net cell proliferation occurs. Reflection on growth regulation *in vivo* leads to the suspicion that cells are subject to negative as well as positive growth regulatory factors. Factors may exist that block cell division even in the presence of all essential growth factors. In the paragraphs that follow, we shall consider the nature of several apparent negative growth controlling responses that have been studied *in vitro*.

When plated on plastic culture dishes in a rich medium, 3T3 cells grow to a density of about 4×10^4 cells/cm² and then cease dividing. But SV40-transformed 3T3 cells will grow to ten times this density. Treatment of transformed cells with fluorouridine (FUdR) allows the selection of "density revertants," which stop growing at a saturation density that approaches that of a normal cell (about 7×10^4 cells/cm²). The FUdR-selected cells are also nontumorigenic. Because cessation of growth in 3T3 and FUdR density revertant cultures occurs at about the stage when cell surfaces come into continual contact with each other, density-dependent growth inhibition is frequently referred to as "contact inhibition." This terminology, however, may be misleading. It is not known whether growth inhibition results from direct cell–cell contacts or from depletion of essential serum growth factors.

Normal cells usually cannot form colonies on top of a nongrowing monolayer culture of other normal cells. By contrast, oncogenic cells frequently form colonies on such monolayers. In an exception to this rule, polyoma-transformed BHK-21 fibroblasts (Py-BHK) are growth-inhibited by contact with normal mouse cells. Failure to grow is not due to an inability of the Py-BHK cells to adhere to the monolayer; rather, DNA synthesis in the hamster cells is markedly suppressed, as shown by autoradiographic measurements. The low serum requirement of these cells suggests that the growth inhibitory effect is not a consequence of medium depletion by the normal cells. When Py-BHK cells are plated onto thin filters, which are subsequently laid on top of mouse monolayer cultures, they grow as well as Py-BHK cells plated on plastic. Moreover, there is a degree of specificity in this type of growth inhibition; monolayers of avian cells do not inhibit growth of Py-BHK cells. These data suggest that (a) inhibition of growth is mediated by a negative signal; (b) transmission of this growth inhibitory signal requires membrane–membrane contact with or close approximation to normal cells; and (c) many cancer cells have lost the capacity to respond to this growth inhibitory signal. In spite of this evidence, it must be emphasized that the argument for a negative regulatory response is not conclusive, and it is possible that the phenomenon of contact inhibition of growth reflects a local depletion of serum factors.

In general, cells in culture may grow at a maximal rate, in which case the expression of differentiated functions is suppressed, or they may be cultured under conditions that result in depressed growth rates and the concomitant appearance of differentiated characteristics. A variety of experimental evidence suggests that cyclic AMP is involved in the negative control of cell growth as well as the positive control of differentiation specific gene expression. Generation times of a large number of tissue culture lines vary inversely with intracellular cyclic AMP concentrations, and growth rates can be slowed by addition of cyclic AMP to the culture medium. Extensive studies with 3T3 cells have shown that the cyclic AMP content of these cells is low during log phase growth but is dramatically elevated when the cells reach confluency, and growth ceases. Further, when confluent 3T3 cells are induced to divide by addition of fresh serum, cyclic AMP levels decline rapidly as growth begins anew.

Intracellular cyclic AMP levels and growth are normally subject to hormonal regulation. For example, a minimal deviation adrenal cortical tissue culture line responds to adrenocorticotropic hormone (ACTH) with increased cyclic AMP content, induction of steroid hormone synthesis, and cessation of cell division. It would appear that animal cells have evolved a complex cyclic AMP-mediated mechanism by which external stimuli at the cell surface control cell proliferation. Since the cyclic AMP synthetic enzyme, adenylate cyclase, is an integral plasma membrane component, a direct interaction between a cell surface receptor and the enzyme can be postulated.

Changes in Membrane Structure and Function Associated with Transformation

In the previous sections, it was suggested that changes in cell growth patterns that occur upon transformation may be due to modifications of membrane structure and function. But, no evidence for such modifications was presented. Numerous studies suggest that transformation is frequently accompanied by quantitative and qualitative changes in the glycolipid and glycoprotein composition of the plasma membranes of the cells. Moreover, plant agglutinins, such as Concanavalin A (Con A), agglutinate and kill transformed, but not normal cells. These agglutinating proteins of plant origin function by binding to sugar residues associated with the glycoprotein and glycolipid constituents of the surface membrane. Temperature-sensitive transformants were agglutinable at the permissive but not at the restrictive temperature, showing that altered sensitivity to the agglutinins was truly associated with the transformed phenotype. Electron microscopic analyses using ferritin-conjugated plant agglutinins revealed that the preferential agglutinability of transformed cells relative to their normal counterparts is apparently due, at least in part, to a change in the mobility of the lectin-binding sites within the membrane matrix. Only in transformed cells did the binding sites appear capable of appreciable lateral diffusion within the mem-

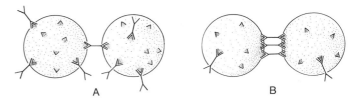

FIGURE 10.1. (A) Normal cells. Receptors for plant agglutinins are available but relatively immobile in the membrane matrix. Few intercellular bridges are made by the bivalent plant agglutinins and the strength of the isolated bridges is inadequate for agglutination. (B) Transformed cells. Quantitatively, the same number of agglutinin binding sites are available, but these sites are mobile within the membrane. Local regions of high binding site concentration develop as a result of intercellular interactions so that the strength of the intercellular bridges that develop within these regions is adequate for agglutination.

brane. Consequently, only in these cells could local concentrations of lectin-binding sites reach a sufficiently high value to permit effective cell agglutination (see Figure 10.1). Thus, a biochemical modification in membrane composition, which occurs as a consequence of transformation, may be sufficient in magnitude to produce dramatic changes in the fluid properties of the cell surface.

Microbial Systems for Studying the Molecular Basis of Neoplastic Transformation

It was noted above that SV40 and other animal viruses are capable of converting normal cells into cancerous cells in a process that requires the integration of viral nucleic acid into the host chromosome. The viral DNA apparently codes for cell surface antigens that can be detected by immunologic techniques and that may either accompany, or confer upon the transformed cell, altered growth regulatory properties. In a superficially similar fashion, the bacteriophage ε-15 is known to alter the structure of the bacterial host surface lipopolysaccharide and thereby change its antigenic properties when the phage DNA is integrated in the bacterial chromosome (lysogeny). Lysogenic infection alters the phage receptor so that ε-15 phage cannot bind to the bacterial surface, although another phage, ε-34, becomes capable of doing so. Two changes in the lipopolysaccharide structure have been identified: acetyl substituents, which normally derivitize galactosyl residues in the polymer, are absent, and the galactosyl residues, which are normally linked to mannosyl residues via α-glycosidic bonds, are changed to the β configuration. Changes in the receptor polysaccharide appear to be due to specific phage gene products: one phage product blocks the synthesis of the bacterial acetylating enzyme; another inhibits the activity of the bacterial α-polymerizing enzyme; and a third is a new enzyme, a β-polymerase, which replaces the inactivated α-polymerase. As a consequence of the action of these three phage proteins, lysogeny changes the structure of the surface lipopolysaccharide, alters the antigenic properties of the cell surface, and abolishes sensitivity of the bacterium to further infection by ε-15 phage.

Animal cells maintained in culture frequently exhibit the property of genetic instability. Specific traits, whether they be cell surface antigenicity, intracellular enzyme activities, or cellular morphologic features may spontaneously be expressed or lost reversibly. The rates at which these differentiated properties are turned on and off are frequently too high to be accounted for by mutation, and several studies now substantiate the conclusion that stable shifts in phenotypic expression need not be due to acquisition or loss of genetic information. Although the molecular basis of these phenotypic shifts has not been determined, the phenomenon, referred to as "epigenetics" or "paramutation," has been attributed to interactions between genetic regulatory alleles.

Many investigators believe that paramutation underlies the pleiotrophic phenotypic changes that accompany viral or chemical transformation of an animal cell. In view of this possibility, it is encouraging that genetic instability has been observed in plants, lower eukaryotes, and bacteria. In several instances, the phenomenon has been extensively studied but in no case is the molecular basis understood. Antigens on the surfaces of bacterial cells have been shown to undergo variations with high frequency. For example, the two flagellar H antigens (H_1 and H_2) of *Salmonella* strains can be alternately expressed, but the two are never expressed simultaneously by a single cell. The two antigens represent alternative flagellar proteins, each of which is coded for by a distinct structural gene. In some bacteria, only the H_1 or the H_2 antigen is expressed; in other strains, these two forms alternate with a frequency as high as 10^{-3}/cell/generation. The frequency at which the transition from one form to the other occurs is determined by a regulatory gene located on the bacterial chromosome adjacent to the H_2 structural gene. It is possible that the H_2 gene arose from a precursor H_1 gene as a result of gene duplication and genetic translocation followed by evolutionary divergence, but the nature of the neighboring regulatory gene (or genes) remains a mystery. It seems clear that an understanding of complex regulatory phenomena in higher organisms will require extensive investigations to elucidate the genetic and biochemical properties of simple bacteria.

Mammalian Growth Regulation and Membrane Biology

In considering the complex problems of growth regulation and neoplastic transformation, we have seen that our inability to answer fundamental questions arises because of insufficient knowledge on each of the subjects dealt with in previous chapters. Neoplastic transformation resulted in altered membrane composition and fluidity (Chapters 2 and 3), which presumably led to enhanced mobility of surface proteins (Chapter 4). Moreover, the abnormal growth properties of cancerous cells correlated with enhanced rates of transmembrane nutrient transport (Chapter 5) and increased sensitivity to insulin-like serum growth factors (Chapter 8). Both of these characteristics may have resulted from altered sensitivity to either membrane reception or a transmission mechanism (Chapters 6 and 7). Finally we noted that cell–cell contacts (Chapter 9) provide growth inhibitory stimuli to normal but not to transformed cells. It seems clear that an understanding of mammalian pathologic conditions resulting from neoplastic transformation will require a more extensive knowledge of many aspects of membrane biology.

Selected References

Brink, R. A. "Paramutation," in *Annual Review of Genetics, Vol. 7.* Annual Reviews, Inc., Palo Alto, Calif., 1974, p. 129.

Brooke, R. F. "Growth regulation *in vitro* and the role of serum," in *Structure and Function of Plasma Proteins* (A. C. Allison, ed.). Plenum Press, New York, 1975.

Clarkson, B. and R. Baserga (eds.). *Control of Proliferation in Animal Cells.* Cold Spring Harbor Laboratory, Cold Spring Harbor, New York, 1974.

Iino, T. Genetics and chemistry of bacterial flagella. *Bacteriol. Rev., 33*:454 (1969).

Pollack, R. (ed.). *Readings in Mammalian Cell Culture.* Cold Spring Harbor Laboratory, Cold Spring Harbor, New York, 1974.

Wright, A. and S. Kanegasaki. Molecular aspects of lipopolysaccharides. *Physiol. Rev., 51*:748 (1971).

The truth is not a citadel of certainty to be defended against
error: It is a shady spot where one eats lunch before tramping on.

Lynn White, *Machina ex Deo*

Index